THE
OIL & GAS ENGINEERING
GUIDE

Hervé Baron

2010

Editions TECHNIP 25 rue Ginoux, 75015 PARIS, FRANCE

Disclaimer:

The author has taken due care to provide complete and accurate information in this work. However, neither the Editor nor the author guarantees the accuracy or completeness of any information published herein. Neither the Editor nor the author shall be held responsible for any errors, omissions, or damages arising out of use of this information.

© 2010, Editions Technip, Paris

All rights reserved. No part of this publication may be reproduced or transmitted in any form or by any means, electronic or mechanical, including photocopy, recording, or any information storage and retrieval system, without the prior written permission of the publisher.

ISBN 978-2-7108-0945-6

Preface

When the author, an alumni of IFP School, contacted me in 2008 to express his desire to share his experience with students, I readily accepted. Such is the practice of IFP School: to involve working professionals among its faculty, to bridge the gap between the academic and professional worlds. This transfer of living knowledge is the foundation of our curriculum.

What I did not expect, though, was that he would take this opportunity to write a book!

What started as a one-day lecture gave rise to a book summarizing 15 years of professional experience.

In doing so, I know that the author took a great deal of pleasure. Not only in the sense that he was producing something useful, but also in the consolidation of the various aspects of his experience as well.

Writing this book was an opportunity for him to stand back from details and derive principles. This resulted in a clear synthesis offering an overall view to the newcomer.

I wish this work great success. I also wish the author to grow in his teaching skills, capturing the audience as he does by covering the subject matter simply as a recollection of first hand experiences.

Jean-Luc Karnik, Dean, IFP School

Acknowledgments

The author wishes to address a special thank to:
- Jean-Luc Karnik, Dean of IFP school, for the opportunity he offered him to give lectures there, which is how this work originated,
- Michel Angot, for his review and significant contribution to this work.

The author also wishes to thank his many colleagues who helped him put this work together:

Luis Marques, Yvonne Parle, Joseph Hubschwerlin, Françoise Penven-L'her, Marie Omon, Bruno Lequime, Frank Sentier, Benoît Le Bart, Claude Baron, Paul Martinez, Didier Salome, Julien Gamelin, Philippe Sainte-Foi, Mirza Sakic, Sonia Ben-Jemia, P. Balasamy Venkatesan, Nedim Tiric, Jacques Félix, Christophe Antoine, Taha Jemaa, Patrice Bois, Tony Rizk Al Gharib, Xavier Carcaud, Roland Bark, Patrick Valchéra, Philippe Defrenne, Gérard Brunet, Gauthier Hatt, Pierre Benard, Erik DuBoullay, Gilles Lumé, Charles Michel, Alain Nicolas, Thierry Lavelle, Séverin Chapuis, Cyndi Cecillon, Christophe Le Cloarec, Yvon Jestin, Christophe Pagé, Alain Guillemin, Bruno Le Guhennec, David Gravier, Sébastien Rimbert, Sergio Sarti, Etienne Karmuta, Christophe Alcouffe, Denis C. Amon, Alexandre Cuénin, Serjun V. Palencia, Bernard Caussanel and Philippe Zuelgaray

The engineering documents shown herein come from real projects executed by Saipem and Technip. The author wishes to thank these two companies for their authorization to publish these documents.

Finally, the author wishes to thank Groupe H. Labbe (www.labbe-france.fr), for their authorization to reproduce pictures of equipment manufactured in their premises.

Reader's feed back:

Not being a specialist of all subject matters described in this work, the author submitted its various sections to the review of senior engineers in each discipline. This will not have removed all errors or avoided omissions. The author will be grateful to the reader that will point them out, comment on the book or make any suggestion for improvement.

Correspondence to the author shall be addressed to: oilandgasengineering@gmail.com.

Table of Contents

Preface		III
Acknowledgments		IV
Usual Engineering Abbreviations		VI
Introduction		1
1.	Project Engineering	5
2.	Getting started	13
3.	Process	15
4.	Equipment/Mechanical	31
5.	Plant layout	41
6.	Health, Safety & Environment (HSE)	51
7.	Civil engineering	71
8.	Material & Corrosion	93
9.	Piping	99
10.	Plant model	123
11.	Instrumentation and Control	131
12.	Electrical	155
13.	Field Engineering	173
14.	The challenges: matching the construction schedule	177
15.	The challenges: controlling information	187
Index: Common Engineering Documents		197
Appendix: Typical engineering schedule		201

Usual Engineering Abbreviations

3D	3 Dimensions
CWI	Civil Works Installation drawing
EPC	Engineering, Procurement and Construction
ESD	Emergency Shut Down
Ex	Explosion protection
FEED	Front End Engineering Design
F&G	Fire and Gas
HAZOP	HAZard and OPerability study
HVAC	Heating, Ventilation and Air Conditioning
HSE	Health, Safety and Environment
IFA	Issue For Approval
IFC	Issue For Construction
IFD	Issue For Design
IFR	Issue For Review
ISO	piping Isometric drawing
MTO	Material Take-Off
PFD	Process Flow Diagram
P&ID	Piping & Instrumentation Diagram
PCS	Process Control System
QRA	Quantitative Risk Analysis

Introduction

The execution of a turn-key Project for an industrial facility consists of three main activities: Engineering, Procurement and Construction, which are followed by Commissioning and Start-Up.

Engineering designs the facilities, produces the list, specifications and data sheets of all equipment and materials, and issues all drawings required to erect them at the construction site.

Procurement purchases all equipment and materials based on the lists and specifications prepared by Engineering.

Finally, Construction installs all equipment and materials purchased by Procurement as per the erection drawings produced by Engineering.

Engineering design is the first, and most critical part, of the execution of a Project. It is indeed engineering that writes the music that will then be played by all project functions: Procurement procures equipment/material as specified by Engineering, Construction erects as shown on engineering drawings.

Engineering is the task of translating a set of functional requirements into a full set of drawings and specifications depicting every detail of a facility.

Engineering involves a variety of specialities, which include Process, Safety, Civil, Electrical, Instrumentation & Control to name a few, and a large number of tasks, from high level conceptual ones to the production of very detailed fabrication and installation drawings.

Cost pressures in the past decade have resulted in the transfer a number of tasks from high cost countries to low cost centres. This does not make it easy for today's engineers entering an Engineering and Construction Contractor to get an overall view of Engineering activities.

Introduction

This work's purpose is to meet this need. It describes in a synthetic yet exhaustive way all activities carried out during the **Engineering** of Oil & Gas facilities, such as refineries, oil platforms, chemical plants, etc.

The work stays on the level of principles rather than going into detailed practices.

In this sense, what is described here is not the practice of a particular Engineering Company but is, to a very large extent, common to all and can be found on any Project.

All illustrative documents (drawings, diagrams, text documents) are actual Engineering deliverables which were used on executed Projects.

Chapter 1

Project Engineering

Engineering of a facility is done in two different steps, a conceptual one, the Basic Engineering, and an execution one, the Detailed Engineering. These two steps are almost always done by different contractors. The Basic Engineering is usually done under an engineering services contract while detailed design is normally part of the facility's Engineering, Procurement and Construction (EPC), also called turn-key, contract.

The scope of **Basic Engineering**, also called **Front End Engineering Design (FEED)**, is to define the facility at a conceptual rather than a detailed level. It entails defining the process scheme, the main equipment, the overall plot plan, the architecture of systems, etc. Basic Engineering stops with the issue of the main documents defining the plant, mainly the Piping and Instrumentation Diagrams, the overall layout of the plant (plot plant), the specification of the main equipment, the Electrical distribution diagram and the Process Control system architecture drawing.

The basic engineering documents serve as the technical part of the call for tender for turn-key execution of the project.

Detailed Engineering takes place during the actual project execution phase. It consists of producing all documents necessary to purchase and erect all plant equipment. It therefore entails producing the specification and bill of all quantities for all equipment and materials. It also entails producing all detailed installation drawings.

Project Engineering

Detailed Engineering integrates vendor information (actual equipment data after design by vendor), as purchasing of equipment actually takes place in this phase, whereas Basic Engineering takes place ahead of equipment purchasing.

The depth of details required in installation related Engineering activities, such as Civil, Steel Structure, Piping, Electrical and Instrumentation, etc. will depend on the split of responsibility that has been agreed with the construction contractor.

It is very common, for instance, that pipes with a diameter below 2" are excluded from the Engineer's scope. Their routing, the material take-off and the procurement of the associated material are under the responsibility of the construction contractor.

The **Engineering Execution Plan** contains a split of responsibility matrix who defines who does what between the Engineer and the Construction Contractor.

		RESPONSIBILITY MATRIX				
ACTIVITY	**STUDY / EXPERTISE ACTIVITY**	**ENGINEERING / PROCUREMENT / SUPPLY**				
		STUDIES	REQUISITIONS	STUDIES	MTO's	SUPPLY
		ENGINEER	ENGINEER	SHIPYARD	SHIPYARD	SHIPYARD
PIPING						
	Plot Plan and Equipment general lay out	X				
UTILITIES PIPING	Utilities upstream of modules			X	X	X
	Utilities Headers inside modules lay out	X			X	X
	Utilities Headers Prelim. Weight report	X				
2" and above	Utilities Headers inside modules drwgs	X				
below 2"	Utilities smaller lines inside modules lay out			X	X	X
	Utilities MTO's inside modules			X	X	X
	Weight report	R		X		
PIPING CLASS	900# Piping Class Specification	X				
	900# Valves, Relief Valves Specifications	X				
	All other Piping Class Specifications	R		X		

X – Responsible
R – Review / Comment

Additionally, Engineering tasks can be distributed between Engineering centres in different parts of the world.

Engineering is split into various disciplines, the main ones being shown on the chart here.

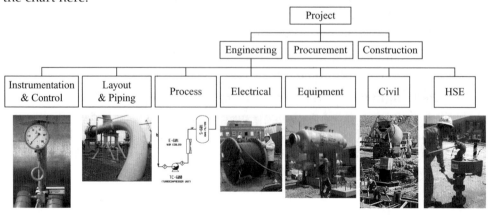

The disciplines are coordinated by the project engineering manager who, like an architect for a building project, ensures consistency between the trades.

Engineering activities are of a various nature. Some disciplines, such as Process, are not much concerned by the geographical layout of the plant: They only produce diagrams (representation of a concept) and do not produce drawings (scaled geographical representation of the physical plant).

Other disciplines are very much concernend with these physical drawings, as shown on the matrix below.

Activity	Engineering Discipline					
	Process	Equipment	Civil	Piping	Instrumentation & control	Electrical
Diagrams	X				X	X
Geographical drawings			X	X	X	X
Architecture drawings					X	X
Calculations	X	X	X	X		X
Equipment or material specification, data sheet & requisition		X		X	X	X
Site works specification			X	X	X	X

Thousands of documents and drawings are issued by Engineering on a typical Project.

These documents can nevertheless be grouped in categories. For instance, although Piping issues as many large scale drawings as required to cover the whole plant area, all are of the same type: "Piping General Arrangement Drawing".

All commonly issued engineering documents are listed in the Index at the end of this work. An example of each one is included in the corresponding discipline section.

There are many inter-dependencies between these documents. For instance, piping routing drawings are issued after the process diagram is defined, etc. These inter-dependencies will be described in the schedule section.

The typical schedule of issue of engineering documents is shown in Appendix. A given document will usually be issued several times, at different stages. Typically, a document is first issued for internal review (IFR) of the other disciplines, then to the client for approval (IFA), then for design (IFD) and ultimately, for construction (IFC).

Most of the documents will also undergo revisions to incorporate the necessary changes or additional details as the design progresses.

A document numbering system is put in place. Document numbers include, besides a serial number, discipline and document type codes. This allows quick identification of the issuing discipline and nature of document.

An **Engineering Document Register** is maintained to show at any time the list and current revision of all documents.

Engineering document register

Document number			Document title	Document revision
A	1	48104	Service building instrument. rooms cables routing	B
A	2	48102	Trouble shooting diagrams	D
A	3	48134	F&G system architecture drawing	E
A	4	50100	Instrument index	B
A	7	50003	Spec for instrument installation works and service	C
A	8	50960	Instrument Data sheets for temperature switches	B
A	9	50110	Requisition for pressure relief valves	B
M	1	62059	General plot plan	B
M	2	62020	Piping details standard	C
M	2	62070	Piping general arrangement Area 1	D
M	4	60100	Special items list	D
M	5	62250	Piping isometrics booklet	C
M	6	60000	Pipes and fittings thickness calculation	A
M	6	62351	Calculation note CN1 - piping stress analysis	A
M	7	60001	General piping specification	C
M	8	60103	Data sheets for station piping material	B
M	9	60200	Requisition for pipes	F

Discipline code	
A	Instrumentation & Control
C	Civil engineering
E	Electrical
G	Project general documents
J	Mechanical
K	Safety
M	Piping & Layout
P	Processes
S	Steel Structures
V	Vessels – Heat exchangers
W	Materials – Welding

Document code	
1	Installation drawings
2	Detail drawings
3	Diagrams
4	Lists – Bill of Quantities
5	Isometrics
6	Calculation notes
7	Specifications
8	Data sheets
9	Requisitions

Drawings are mainly of 3 types, as follows:
- diagrams, such as Piping & Instrumentation Diagrams, which show a concept,
- drawings, such as Piping General Arrangements Drawings, which show a scaled geographical representation of the plant. The representation may be a plan (top), an elevation (front, side, back) or a cross section view.
- key plans, which shows the division of the plant territory in multiple drawings of a particular type, covering all plant areas at high enough a scale.

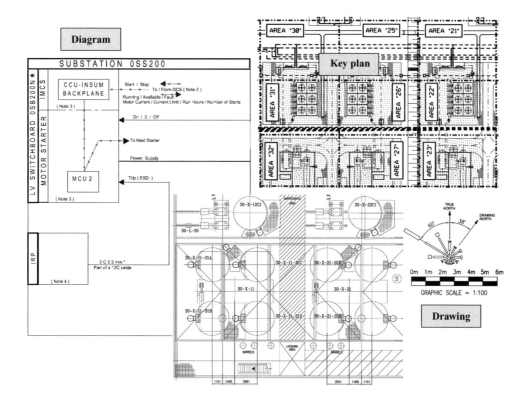

A3 is the common size for diagrams and A1 for drawings (in order to cover the maximum surface area in the later case). A0 is not often used as it does not easily unfold at the job Site.

Overall drawings, such as the overall plot plan are issued with a scale of 1/500 or 1/100, depending on size of plant, while detailed installation drawings issued by Piping and Civil for each plant are issued with a 1/50 scale.

Project Engineering

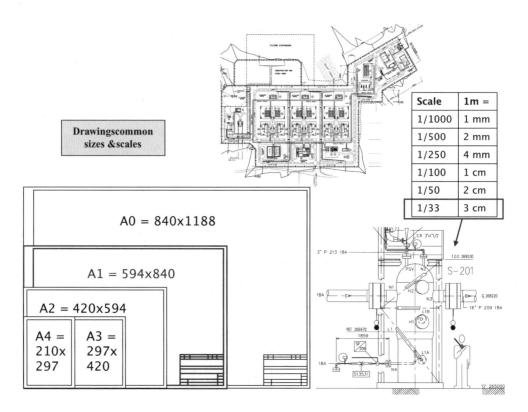

Engineering is the integrator of vendor supplied equipment. Such integration is highly dependant on information from the vendors (size of equipment, power consumption, etc.). One of the challenges faced by Engineering is the management of such an integration, which requires timely input of vendor information not to delay design development.

The plant owner is involved in the Engineering process as they need to review and approve the design and check the compliance to their requirements.

The main information flows are depicted in the diagram that follows.

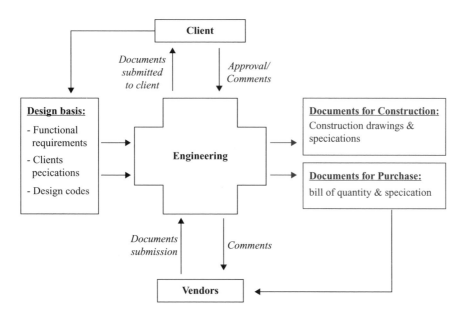

Managing the flow of information at the interfaces (between disciplines and with vendors) is highly critical, as will be explained in the conclusion to this work.

Chapter 2

Getting started

Design of a new process facility starts by the definition, as per its owner requirements, of its *function*. In short, what is the process to be performed: liquefaction of natural gas, separation and stabilisation of crude oil, etc., the required capacity, the feed stock and products specifications and the performance (availability, thermal efficiency, etc.).

A typical duty for an oil platform would be:

> "The facilities will be designed to handle production rates of 200 kbpd (annual average) of oil production and a peak of 15 Msm3/d of gas production.
>
> The full wellstream production from the subsea wells will be separated into oil, water, and gas phases in a three-stage flash separation process with inter-stage cooling designed to produce a stabilized crude product of 0.9 bara true vapor pressure. Water will be removed in the flash separation/ stabilization process in order to reach of 0.5 vol.% BS&W oil specification. The produced gas will be compressed, dehydrated and be injected into the reservoir to maintain pressure as well as conserve the gas."

On top of the functional requirements, come a number of Client specifications, technical requirements for equipment for instance, that will ensure that the facility will last and have the required availability. For instance, design and mechanical requirements will be specified for pumps, so as to limit wear and need for maintenance, to ensure uninterrupted operation over a specified service life.

The Client requirements are found in the Contract Engineering Basis, which describes "what" and "how" to deliver from a technical perspective.

It includes both project specific functional requirements, which include the scope of work and the design basis, and general requirements, such as Client's design standards and specifications.

In order for all engineering disciplines to work with a concise document summarizing the main design bases, the Engineering Manager issues the **Engineering Design Basis** document which gives reference information such as feedstock composition, environmental data, performance requirements, applicable specifications, etc.

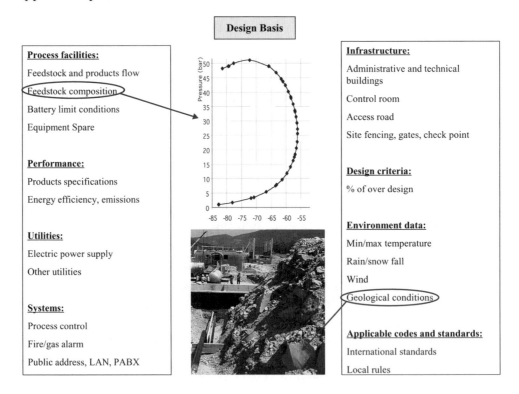

Chapter 3

Process

◆

How to process the inlet fluid(s) to produce the required product(s), i.e., the Process, will be defined by performing **Process simulations**. These simulations use thermodynamic models to simulate fluid behaviours under the different process operations: phase separation, compression, heat exchange, expansion, etc.

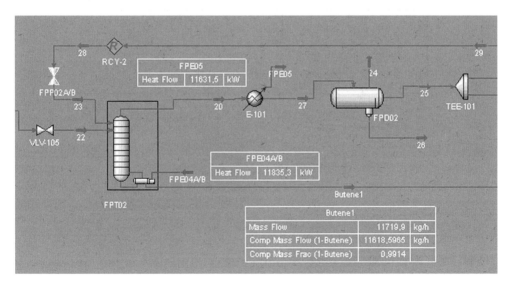

The software also calculates the duty of the equipment. For instance, the software will calculate the required capacity of the compressor to bring gas of such composition from a given inlet pressure to a given discharge pressure. Such calculation is difficult to make manually as petroleum fluids contain a large variety of components. The software incorporates thermodynamic models, which include the properties of all these components, and calculates the difference in enthalpy between the compressed and non-compressed gas, hence the required compressor capacity.

Such mechanical duty will in turn determine the size of the driver (gas turbine for instance) by applying typical compressor efficiency, losses in gear box, etc.

Various schemes are simulated to find the optimum process scheme. This optimisation is done to match a few constraints, e.g., the ratio of outlet to inlet pressure in a gas compressor is around 2.5 (above that the gas temperature becomes too high), gas turbines come in a range of stepped – not continuous – sizes, e.g., 3/5/9/15/25 MW.

Running process simulations will allow to try several scenarios and optimize the process, i.e., reduce the number of equipment, energy consumption, etc.

In the case of crude oil stabilisation for instance, pressure levels will be optimized, in order to reduce the number of separators while keeping pressures at values that allow easy gas re-compression (for injection back into the reservoir).

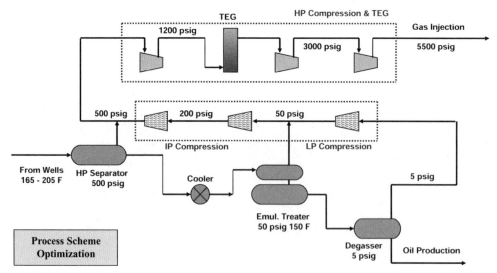

Once the optimum scheme is found, Process displays it on the **Process Flow Diagrams (PFDs)** that show the process equipment, e.g., separator, heat exchanger, compressor, etc., and their sequence.

Process

Process Flow Diagram (PFD)

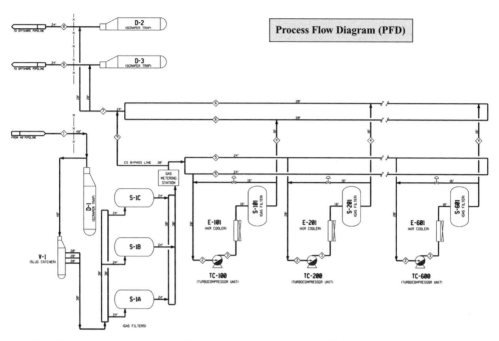

The Process **Equipment list** is the register of all Process equipment. It is derived from the PFDs.

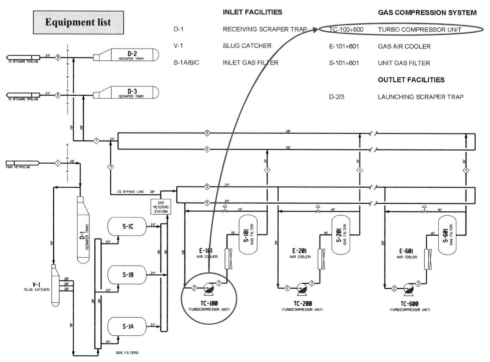

Process

Process simulations are run for all operating cases, such as initial year of operation, plateau level, operating case at field end of life when water to oil ratio has increased significantly, etc.

This determines the required capacity of each equipment. Minimum and maximum duties are identified covering the full range of operating cases. The range of operating conditions for each line (pressure, temperature, flow) is also identified, which will allow adequate line specification and sizing.

The results are tabulated in the **heat and material balance**, which shows the flow, composition and condition of each stream.

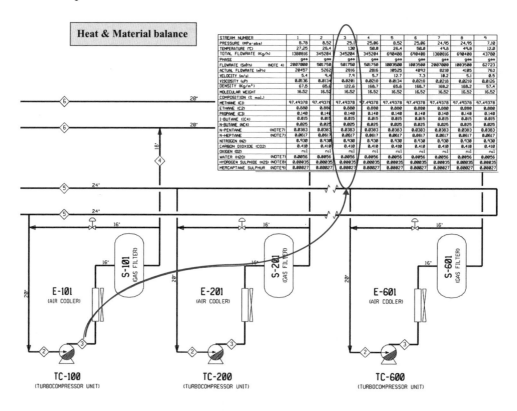

Process then performs the **sizing of equipment** as per the required process duty, e.g., size of gas compressor according to required gas flow and gas gravity, size of cooling water pump as per process fluid cooling requirements and calorific value, etc. and Site conditions, e.g., temperature of available cooling medium (air/sea water).

Process

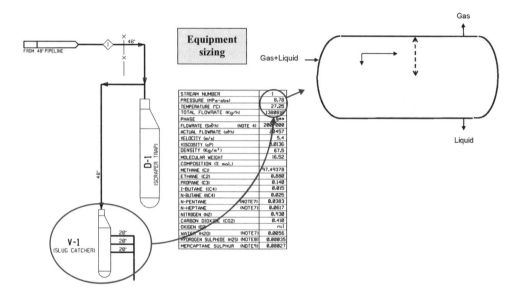

The process duty of each equipment is specified in a **Process data sheet**.

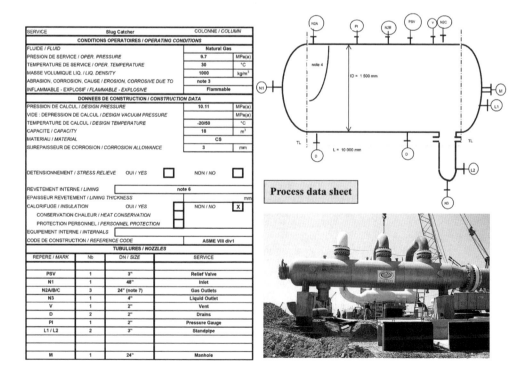

Process data sheet

The PFDs also show how the process will be controlled, by indicating the required controls (of pressure/temperature/flow) throughout the process scheme. This is further described by Process in the **Process Description and Operating Philosophy**.

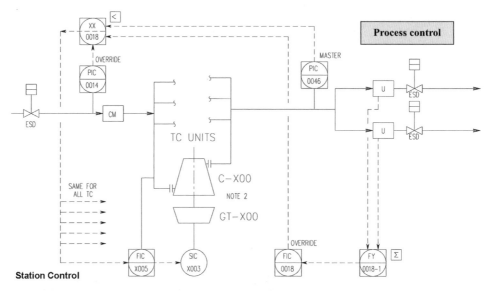

Station Control

Control of the station shall be carried out, manually or automatically, by adjusting the revolutions of the units, controlling the most critical of the following parameters:

- suction gas pressure (override);
- discharge gas pressure (master);
- gas flow rate (override).

As the process diagram is further detailed, PFDs are translated into **Piping and Instrumentation Diagrams (P&IDs)**.

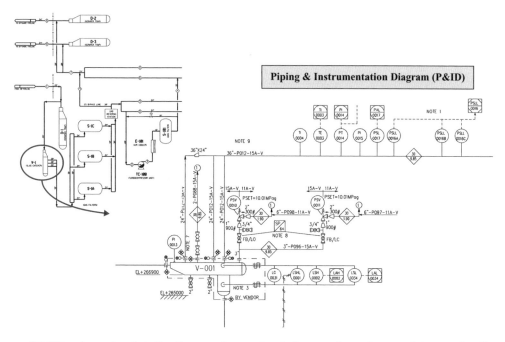

P&IDs show in details the equipment, piping, valves (manual/motorized), instrumentation, process control and emergency shutdown devices.

P&IDs do not only include all lines, instruments, valves required during normal operation, but also the ones required for maintenance, plant start-up and all operating cases.

They will include, for instance, equipment isolation valves, depressurization and drainage lines. They will also include a recycle line required for operation of the plant at low throughput, etc.

The **Legend and Symbols P&ID** shows the meaning of the graphical elements and symbols used on the P&IDs. For instrumentation, an international symbols and identification standard is generally used, providing a means of communicating instrumentation, automation and control requirements that all parties can readily understand.

22 Process

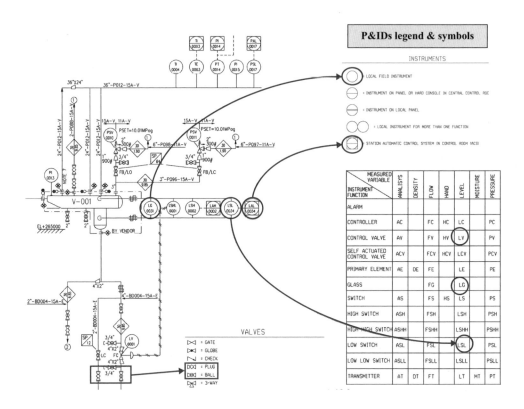

The P&IDs are developed by Process as per the various Operating, Safety and Maintenance requirements:

- Equipment isolation philosophy: valves and bypass to be provided,
- Requirements for start-up and shutdown, i.e., additional bypass/pressurization, drain lines, etc.,
- Equipment sparing/redundancy philosophy, e.g., 2 pumps, each 100%, one operating and one spare,
- Process controls, which are directly shown on P&IDs by means of dotted lines between controlled process parameter (pressure, flow, temperature) and control valve. Process automations (ON/OFF controls) are described in specific diagrams called **Process Cause & Effects Diagrams**.

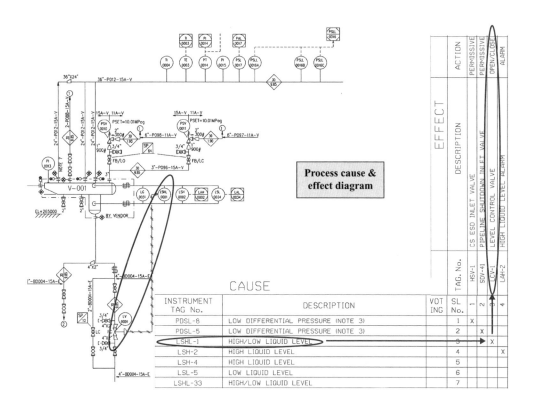

- Process safety automations: sensors initiate process shutdown in case of upset of process parameters. Their detailed logic of operation is shown on the **Emergency Shutdown (ESD) Cause & Effect Diagrams**.

Process

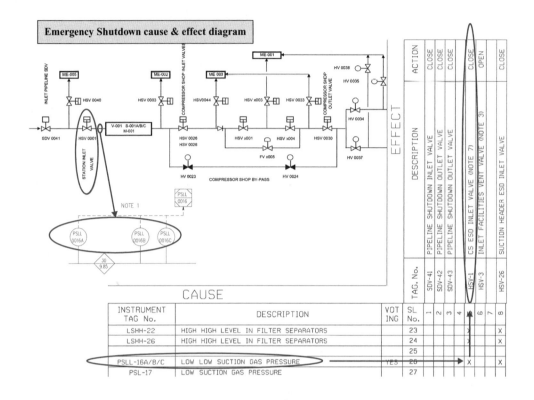

Emergency Shutdown cause & effect diagram

- Plant emergency isolation and depressurization requirements. To ensure it can be returned to a safe condition in case of emergency, the plant is split into sections that can be isolated from each other. Such isolation is achieved by means of emergency shutdown valves (ESDVs). Each section can also  be depressurized. The split into sections, which determines the number of isolation and depressurisation valves to be provided and their location, is shown on the **ESD simplified diagram**.

Process

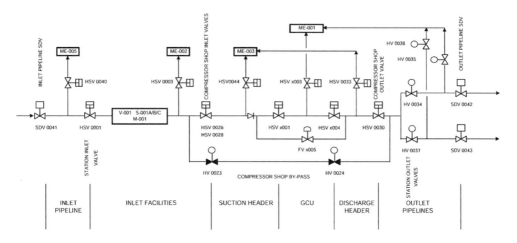

Besides the individual process safety automations described above, Process designs the system to safely bring the plant to a stand still in an emergency. This is done in various degrees, called ESD levels, from the shutdown of a local process unit, to the shutdown or even shutdown and depressurisation of all facilities.

The ESD levels are cascaded: the overall plant shutdown initiates each of the individual unit shutdown, as shown on the **ESD logic diagrams**.

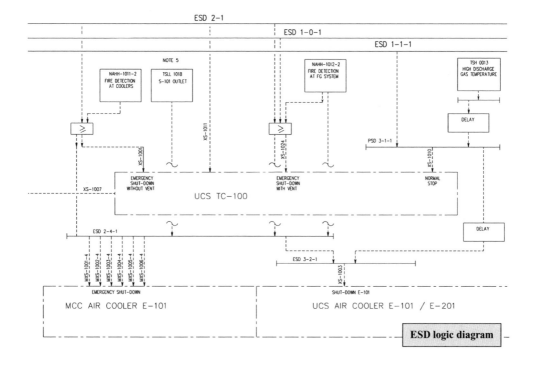

Process

The definition of the levels, the initiating causes and the actions implemented for each one, are described in the **Emergency Shutdown and Depressurisation Philosophy**.

Process discipline is also in charge of designing the relief system. A relief system is used to safely release overpressure in case of process upset, or to completely depressurize the plant in case of emergency such as a major leak, etc. Process designs the relief system: diameter of relief lines, design pressure of liquid collection vessel (flare knock-out drum), capacity of flare tip, etc. to cover all relief scenarios.

Relief system design criteria are given by codes or client requirements, such as the requirement to depressurize the plant to 7 bars in less than 15 minutes in an emergency.

The **Flare Report** details the relief calculations and results, including the levels of low temperature reached in the pressure vessels and relief lines during depressurization. Very fast depressurization from high pressures to very low pressure in a few minutes leads to very low temperature. The depressurization case determines the low design temperature of the pressure vessels and the flare system. It may dictate the use of special materials such as low temperature carbon steel, or even stainless steel.

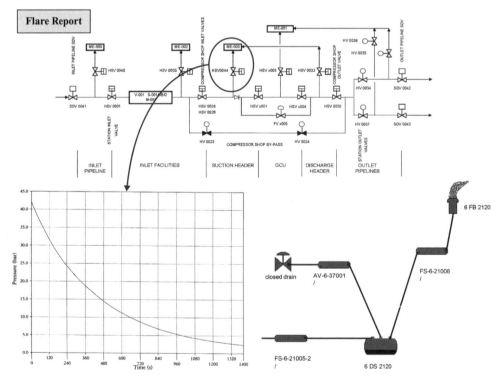

Flare heat radiation calculations are done as part of the flare study, to define the height of the flare stack. The required stack height is the one that gives low enough a level of heat radiation at grade/closest operating areas levels.

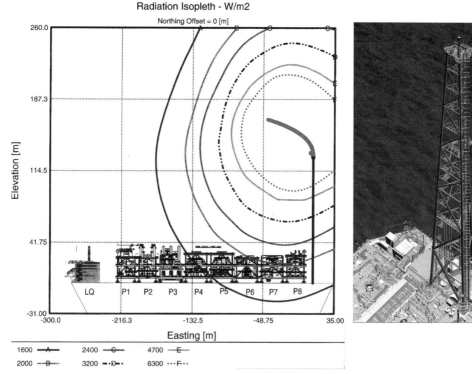

Process

The PIDs are the main vehicle by which the Process design will be shown to the Client, to whom they are issued for approval.

They are also the basis on which Piping, Instrumentation and Control disciplines will develop their design. For instance, they show Instrumentation discipline the detailed requirements: not only the process parameter to be measured (Flow, Pressure, Temperature), but also whether the measure shall be available locally in the field only or on central console in the control room, whether the value must be recorded (to keep history), etc.

P&IDs also record the precise interface with vendor equipment and packages (piping connections, exchanged instrument and control signals, etc.).

The P&IDs are living documents, which are amended with inputs from numerous parties. These inputs are sourced from Client, HAZOP review, all disciplines, Vendors e.g., size of control valves once sized, equipment and packages interface information, etc.

While developing the P&IDs, Process groups the various fluids, based on their operating conditions (pressure, temperature) and corrosiveness in the **Process fluid list**.

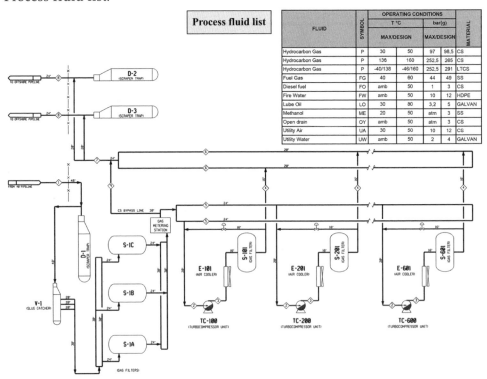

Process fluid list

FLUID	SYMBOL	OPERATING CONDITIONS				MATERIAL
		T °C		bar(g)		
		MAX/DESIGN		MAX/DESIGN		
Hydrocarbon Gas	P	30	50	97	98,5	CS
Hydrocarbon Gas	P	138	160	252,5	265	CS
Hydrocarbon Gas	P	-40/138	-46/160	252,5	291	LTCS
Fuel Gas	FG	40	60	44	49	SS
Diesel fuel	FO	amb	50	1	3	CS
Fire Water	FW	amb	50	10	12	HDPE
Lube Oil	LO	30	80	3,2	5	GALVAN
Methanol	ME	20	50	atm	3	SS
Open drain	OY	amb	50	atm	3	CS
Utility Air	UA	30	50	10	12	CS
Utility Water	UW	amb	50	2	4	GALVAN

Process

Process also numbers each line and maintains the corresponding register, called **Process Line List**. It shows the process information for each line, namely, fluid type, fluid phase, operating and design temperature and pressure, etc.

Line Number			Line Size	Class	P&ID Dwg. No.	Line Connection		Fluid Phase	Operating Condition		Density	Design Condition			Full Vacuum
Fluid Code	Unit Code	Seq No.				From	To		Press	Temp		Press	Temp (Max.)	Temp (Min.)	
									barg	degC	kg/m3	barg	degC	degC	(Y/N)
GN	71	61106	22	3C3AS1	D-80-212	LNG STORAGE	UNIT 93	V	27,6	55	18,2	34,5	100		N
GN	71	61106	20	3C3AS1	D-80-212	LNG STORAGE	UNIT 93	V	27,6	55	18,2	34,5	100		N
GN	71	61106	12	3C3AS1	D-80-212	LNG STORAGE	UNIT 93	V	27,6	55	18,2	34,5	100		N
LNG	71	60001	32	3R0JLL	D-80-302	668-P001 A/B/C	LNG RUNDOWN HEADER	L	11,1	-159	439	30	80	-167	N
LNG	71	60001	22	3R0JLL	D-80-302	668-P001 A/B/C	LNG RUNDOWN HEADER	L	11,1	-159	439	30	80	-167	N
DOW	72	63000	0,75	1P1	D-72-204	72-P061A	DOW	L	0	48	1000	2	82		N
DOW	72	63001	0,75	1P1	D-72-204	72-P061B	DOW	L	0	48	1000	2	82		N
DOW	72	63002	0,75	1P1	D-72-204	72-P062A	DOW	L	0	48	1000	2	82		N
DOW	72	63003	0,75	1P1	D-72-204	72-P062B	DOW	L	0	48	1000	2	82		N

Process calculates the diameter of lines, based on hydraulic requirements (maximum pressure drop allowed or erosion criteria), shows the details of the calculation in **Calculations notes** and indicates the resulting line diameters on the P&IDs.

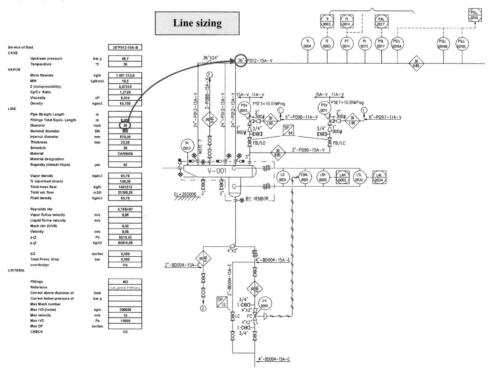

Finally, Process issues the **Operating Manual**, containing a detailed description of the facilities, instructions for start-up, operation and maintenance.

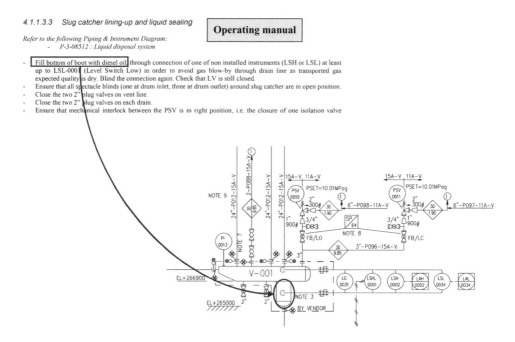

4.1.1.3.3 *Slug catcher lining-up and liquid sealing*

Operating manual

Refer to the following Piping & Instrument Diagram:
- *P-3-08512 : Liquid disposal system*

- Fill bottom of boot with diesel oil through connection of one of non installed instruments (LSH or LSL) at least up to LSL-0001 (Level Switch Low) in order to avoid gas blow-by through drain line as transported gas expected quality is dry. Blind the connection again. Check that LV is still closed.
- Ensure that all spectacle blinds (one at drum inlet, three at drum outlet) around slug catcher are in open position.
- Close the two 2" plug valves on vent line.
- Close the two 2" plug valves on each drain.
- Ensure that mechanical interlock between the PSV is in right position, i.e. the closure of one isolation valve

The operating manual provides reference information such as the capacity of all vessels, set-points of controllers, alarms, safety switches, etc.

TAG	Position	Control device	PID	Unit (1)	Set point	Alarm low	Alarm high	Range (2)	Comment
	Inlet facilities								
PCV0001	Pilot gas for level valves	LVs	8513	bar	11				See note 11 on P-3-08513
LSHL0001	Slug catcher D-001 boot	LV-0001	8512	mm	-150/50				
LAH0002	Slug catcher D-001 boot		8512	mm			200		
LAL0034	Slug catcher D-001 boot		8512	mm		-450			
PIC0014	Header inlet filters separators S-001		8550	bar	67				Act as override only
PAL0017	Inlet gas filters S-001 inlet header		8512	bar		64,5		50 - 70	
LAH0007	Inlet gas filter S-001A upstream		8513	mm			290	-250 to	Level 0 is bottom vessel axis level.

The operating manual contains information about *systems* (process, utility, emergency shutdown) operation. Information on the operation and maintenance of individual equipment are found in the equipment vendor documentation instead.

Chapter 4

Equipment/Mechanical

◆

Equipment, also called Mechanical, discipline includes various specialities, such as pressure vessels, heat exchangers, fired equipment, rotating equipment and packages.

Equipment/Mechanical

Heat exchangers are designed and sized by Engineering, as they are of a standard non-proprietary design (shell and tube, etc.). A computer software is used to model the heat transfer for the specified geometry (number of tubes, position of baffles, etc.). The input and results of the calculation are recorded on a **calculation sheet**.

Rating - Horizontal Multipass Flow TEMA BEU Shell With Single-Segmental Baffles

Process Data		Hot Shellside		Cold Tubeside	
Fluid name		Condensate		Product Gas	
Fluid condition		Sens. Liquid		Sens. Gas	
Total flow rate	(kg/s)	150.817		809.563	
Weight fraction vapor, In/Out	(--)	0.000	0.000	1.000	1.000
Temperature, In/Out	(Deg C)	129.30	39.90	24.40	39.98
Temperature, Average/Skin	(Deg C)	84.6	48.84	32.2	40.15
Wall temperature, Min/Max	(Deg C)	34.07	69.44	33.65	64.65
Pressure, In/Average	(kPa)	500.007	496.549	7530.11	7512.40
Pressure drop, Total/Allowed	(kPa)	6.916	100.000	35.423	50.000
Velocity, Mid/Max allow	(m/s)	0.27		9.19	
Mole fraction inert	(--)				
Average film coef.	(W/m2-K)	751.36		2280.27	
Heat transfer safety factor	(--)	1.000		1.000	
Fouling resistance	(m2-K/W)	0.000200		0.000200	
Overall Performance Data					
Overall coef., Reqd/Clean/Actual	(W/m2-K)	404.25 /	501.27 /	410.71	
Heat duty, Calculated/Specified	(MegaWatts)	32.7979 /			
Effective overall temperature difference	(Deg C)	31.6			
EMTD = (MTD) * (DELTA) * (F/G/H)	(Deg C)	36.51 *	0.8668 *	1.0000	

See Runtime Messages Report for warnings.	
Exchanger Fluid Volumes	
Approximate shellside (L)	3872.5
Approximate tubeside (L)	12521.2

The design and sizing of other equipment, such as compressors, etc. is proprietary and done by their vendors.

The Process data is completed by the Equipment specialist, which produces the **Mechanical data sheet**. The Mechanical Data Sheet specifies additional requirements besides the process duty, such as codes and standards to be applied, performance, energy efficiency, type and materials of construction, etc.

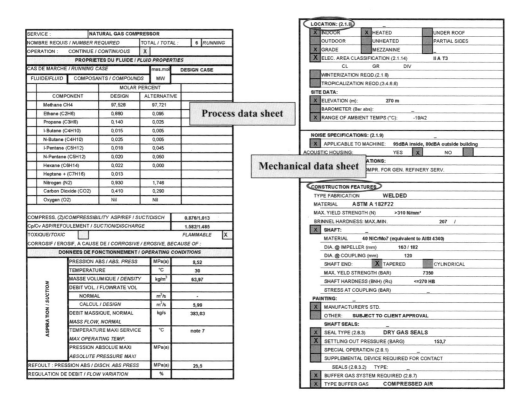

Pressure vessels, on the other hand, are specified in details to the manufacturer: type and position of internals, dimensions, number, size and elevations of nozzles, etc. which come from process requirements. For a gas/oil/water separator for instance, the section of the vessel will be sized to reduce the gas velocity enough to achieve proper gas/liquid separation, the liquid section volume will be defined so as to provide enough residence time to achieve adequate oil/water separation, e.g., 3 minutes for light oil and 8 minutes for heavy oil, etc.

For a distillation column, the distance between trays will be defined by the process licensor, which will in turn set the elevation of the column, nozzles, etc.

The detailed arrangement of the vessel is specified to the manufacturer by means of a **vessel guide drawing**.

Equipment/Mechanical

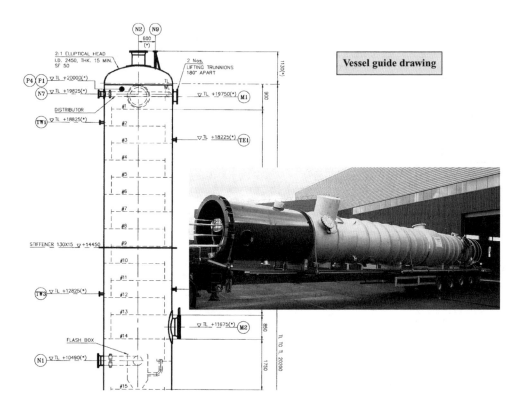

This guide drawing is also issued to Piping for routing of connected pipes. The orientations of nozzles are not defined on the guide drawing. Instead, they are defined by Piping following the piping routing studies.

A **supply specification** is prepared, describing the entire scope and limits of supply and listing all applicable specifications. The piping specification, for instance, should be referenced if the supply includes piping, the electrical specification should be included if the supply includes electrical equipment, etc.

Many pieces of equipment are indeed not purchased on their own, such as a pressure vessel or a heat exchanger, but as a package. A package is a set of equipment, purchased as a functional unit, e.g., a water treatment unit. It comes complete with all equipment, piping, instrumentation, cables, etc. already installed on one or several "skids" (frames). This approach, which consists of purchasing a part of the plant already pre-fabricated, reduces construction time at Site, as assembly is carried out at the vendor's premises instead.

Equipment/Mechanical

The scope of supply of the package vendor in all disciplines must be precisely defined. For a package made of several parts, for instance, the party who is supplying the interconnections (pipes, cables) between the parts must be specified. A detailed matrix, such as the one shown below for Instrumentation, is the most efficient way to precisely define the split of responsibilities and battery limits.

TYPICAL SCOPE OF SUPPLY

C : CONTRACTOR   PC : PACKAGE CONTRACTOR   B.L. : BATTERY LIMITS

DESIGNATION	DESIGN		SUPPLY		INSTALLATION	
	C	PC	C	PC	C	PC
1 feeder 230 VAC, 50 Hz for instrum.	X		X		X	
1 instrument air supply at B.L.	X		X		X	
Electrical distribution		X		X		X
Air distribution		X		X		X
Junction boxes at B.L.		X		X		X
Air connection at B.L.		X		X		X
Interconnecting principles	X	X				
Instruments inside of B.L.		X		X		X
Instruments outside of B.L.	X		X		X	

The specification and the data sheet are attached to a document called a **Material Requisition**, which is the document Engineering issues to Procurement for Purchasing the equipment. The requisition precisely defines the equipment/material to be supplied and the exact scope of supply and services, e.g., what calculations are to be done by the vendor. It also specifies the quality control requirements, the documentation to be supplied by the vendor and its delivery schedule. The documents required from vendors are of different types:

- Study documents, such as P&IDs, calculation notes, general arrangement drawings, etc.

- Interface documents, showing all connections at the supply's battery limits in all disciplines: anchor bolts and loads on foundation, piping connections, electrical and instrumentation connections, etc.

  The interface documents and their timely submission are of primary importance to Engineering, for integration of the equipment/package into the overall plant. Provision of these documents must be synchronised with the engineering schedule. Penalties are specified for late submission of critical documents by the vendor.

- Documents required at the construction site: preservation procedure, list of components (packing list), lifting instructions, commissioning and start-up instructions.

Equipment/Mechanical

- Documents to be retained by the plant owner: manufacturing records, operating and maintenance manual, list, references and drawings of spare parts.

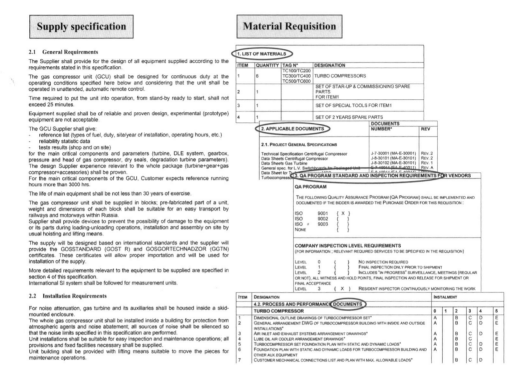

Upon receipt of the inquiry the vendors will perform their own design.

For a compressor, for instance, this will entail defining the number and design of the impellers to match all operating cases with maximum efficiency. The vendor will submit such performance data in their proposal.

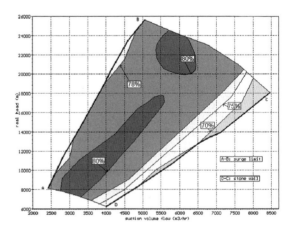

Equipment/Mechanical

Once the bids are received from vendors, technical appraisal is carried out to both confirm compliance to requirements and to compare the offers from the various vendors.

The detailed technical analyses of the bids are shown in the **Technical Bid Tabulation** document. It covers scope of supply and services, compliance to performance guarantees, design and fabrication codes and standards, inspection and quality requirements, supplier's references in similar supplies, etc. For each item, the specified requirements are shown together with what is offered by each vendor.

Technical Bid Tabulation				Requirements	Supplier 1	Supplier 2
1 -SCOPE OF SUPPLY SUMMARY (continued)						
1.6 Piping / Structure / Painting / Misc.						
Piping:						
Interconnecting piping between skids or equipement (fuel, lube oil, water, steam, etc...):				included	included	included
Painting						
Non insulated equipment (motor, valves, steel structures, platforms, etc...)				Max. at shop		up to final coat
Boiler block				Max. at shop	sandblasting & primer + insulation at shop	Primer + insulation/lagging at shop
1 -SCOPE OF SUPPLY (continued)						
1.7 Services						
Shop inspection and tests (as a minimum)					included	
Superheater				As per ASME	Hydrotested before shipping	Hydrotested before shipping
Boiler				As per ASME	Hydrotested before shipping	Hydrotested before shipping
1 -SCOPE OF SUPPLY (continued)						
1.8 Codes & standards						
Boiler presure parts & safety valves				ASME I	ASME I with S stamp	ASME
Pressure parts materials				ASME I	ASME I with S stamp	ASME
2 -OPERATING CONDITIONS						
2.1 Design conditions						
Feed Water Temperature @ BL	(MCR/Peak Load)		°C	120	120	120
Feed Water Pressure	(Mini required / Mecha desi)		barg	68 / 90	65 / 90	Yes
2.2 Guaranteed performances						
Steam Flow	(MCR)		t/h	240	240	240
Steam Outlet Temperature at BL			°C	384 +/- 5	384 +/- 5	384 +/- 6
Steam Outlet Pressure at BL			barg	41,3 +/- 1	41,3 +/-1	41,3
3 - CONSTRUCTION DATA						
3.1 General						
Boiler area dimensions	W x L (w/o eco / w eco)		m	By Vendor	~16000x25000 / ~17000x25000	~24076x27147 / ~25231x26916
Boiler dimensions	W x L x H		m	By Vendor	12500 x 15500 x 10500	8670 x 19000 x 10250
3.2 Steam Drum						
Pressure	operating / design		barg	As per ASME	47,5 / 54,0	tbd / 56,54
Temperature	operating / design		°C	By Vendor	262 / 295	382 / 343
Length (TL-TL)			mm	By Vendor	14200	14833

Equipment/Mechanical

Following this detailed technical analysis, including clarification meetings held with suppliers, the technical acceptability of each bid is advised by Engineering to Procurement.

Once the equipment is purchased, and before proceeding with fabrication, the vendor submits its design documents to Engineering for review and approval. Vendor documents are checked by Engineering for compliance with the purchase order specifications. Comments from the various disciplines are consolidated. The document is returned to the vendor with a code, result of the review, instructing the vendor either to proceed or to revise its design and resubmit it for further review.

	COMMENT STATUS : THE APPROVAL OF THIS DOCUMENT DOES NOT RELIEVE THE SUPPLIER OF ITS CONTRACTUAL RESPONSABILITIES			
1	NO COMMENT OR FORMAL COMMENTS PROCEED WITH FABRICATION RESUBMIT WITH UPPER REVISION STAMPED APPROVED FOR CONSTRUCTION	4	FOR INFORMATION REFERENCE ONLY	
2	APPROVED AS NOTED PROCEED WITH FABRICATION IN ACCORDANCE WITH COMMENTS RESUBMIT CORRECTED DOCUMENTS FOR APPROVAL WITH UPPER REVISION	5	FINAL DOCUMENT	
3	DISAPPROVED DO NOT FABRICATE RESUBMIT CORRECTED DOCUMENT FOR APPROVAL WITH UPPER REVISION	CHECKED BY :		DATE :

Vendor documents provide information on the equipment, such as dimensions, weight, electrical and other utilities consumption, etc. which Engineering incorporates in the overall plant design.

A register is maintained of all equipment: the **Equipment Summary**. Such register is used, for an on-shore project, by the contractor at Site to know how many equipment will have to be installed for its planning purposes and what is the capacity of the cranes required to lift these equipment in order to mobilize the proper cranes.

Equipment/Mechanical

SYSTEM No	ITEM No	Qty of Equip	DESCRIPTION	Design Conditions (gage press.)		Orientation	Vessels Dimensions		P&ID (or Dwg) No.	Unit Est. Shipping Weight	Grade level	MATERIAL note 3
				DP	DT		ID	T-T				
				MPa	°C		mm	mm		ton		
1			**PROCESS**									
1.1			**INLET FACILITIES**									
	D-001	1	Receiving scraper trap	10.01	50	H	1400	9800	P-3-08511	20	261000	CS
	V-001	1	Slug catcher	10.01	50	H	1500	10000	P-3-08512	30	261000	CS
	S-001A/B/C	3	Inlet gas filter	10.01	50	H	1450	3210	P-3-08513	(40)	261000	CS
	M-001	1	Gas metering station package	9.85	50							
	U-060	1	Analyser house	9.85	50							
1.2			**OUTLET FACILITIES**									
	D-002	1	Launching scraper trap	26.5	70	H	750					
	D-003	1	Launching scraper trap	26.5	70	H	750					
	T-050	1	Methanol tank	ATM	50	H	1600					
	P-051A/B/C	3	Methanol injection pump	26.5	50							
	P-052	1	Methanol portable pump	0.07	50							

On an Off-Shore project, the equipment summary helps to prepare the **weight report**.

A **lifting study** is also produced, based on the weight derived from the Material Take-Off in each discipline, to estimate the weight and the centre of gravity. It serves to validate or not the lifting feasibility by the selected crane. In the case where the load exceeds the hook capacity, a weight management is required, to modify the arrangement of the module or to decide to remove a part of the module and reduce the weight for the lifting phase.

Equipment/Mechanical

Weight report	Detail for module X		Reported weight (te)		Center of Gravity		
					East	North	Elevation
Riser protec. / Acc. ladders	Boarding access ladders		1098	171	100,0	222,6	86,0
	Riser Protector			915			
	Cathodic protection			12			
	Mooring Equipment		493	493	97,0	242,1	101,1
Instrumentation & electrical Equipment	Instrumentation equipment		99	13	100	100	87,5
	Electrical Equipment			49			
	Electrical cable integration			32			
	Electrical cable tray / support			5			
Riser inst. winch support and Casings	Paint on riser / caisson		501	14			
	Fire water caisson			66	100	100	83,75
	SW lift caisson			43	66,4	215	83,75
	Suction Hoses			126			
	Riser instalation winch support			252	136,1	NA	102,4
	Total		2190		95,5	249,3	87,9

(*) Gross Estimation of the centre of gravity

Chapter 5

Plant layout

Once the plant equipment is defined, upon completion of the Process Flow Diagrams (PFDs), Plant layout (also called installation) discipline performs installation studies, which consists of defining the topographical organisation of the facility.

An industrial facility is usually split into 3 zones: Process, Utilities and Offsite.

- The process units are where the feedstock is processed into products,
- Utilities units deal with electrical power generation, production and handling of utility fluids such as steam, heating/cooling medium, water, compressed air, nitrogen, etc. and treatment of the waste fluids, such as rain and oily water, drains, waste gas, etc.,
- Offsites consist of product storage, shipping facilities and of buildings.

An Off-Shore facility will also include living quarters (LQ) and a helicopter landing pad, located as far as possible from the process units.

The site where the plant is to be built will impact its layout. A restricted land plot size will drive a vertical stacking rather than an horizontal spread of the plant equipment, a sloped relief will decide a terraced arrangement to minimize the earthworks, uneven soil geotechnical properties will impose constraints for location of heavy or critical installations (large storage tanks, turbo-machinery, etc.).

Plant layout

The plant layout takes into account the plant environment: location of access/exit roads, external connecting networks: pipelines, electrical grid, water supply, etc. It is depicted on the **General Plot Plan**, which is the base graphical document used to locate all items of equipment, structures, buildings, roads and boundaries for the overall plant complex.

The location of the various units, and that of equipment within units, is determined following a number of principles, primarily related to safety.

Hazardous units, such as gas compression units, are located far away from vital units, such as power generation, and manned areas, such as living quarters.

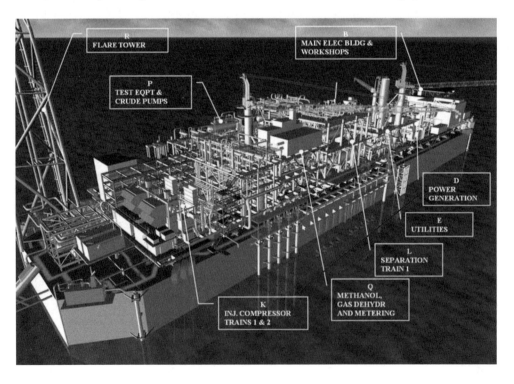

Plant layout

43

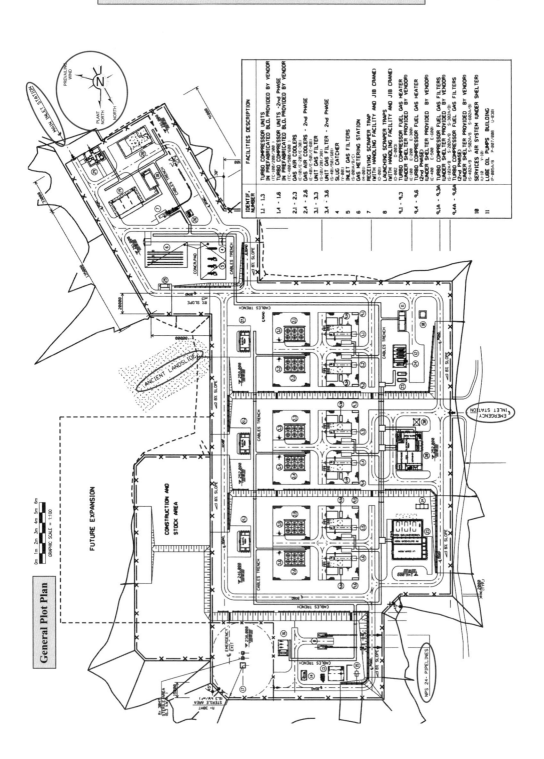

General Plot Plan

Plant layout

Units are classified in terms of risk of releasing flammable materials (leak) or igniting them (explosion, fire). The risk level mainly derives from operating conditions: the higher the pressure and the temperature, the higher the risk.

Risk are classified in High (HH), Intermediate (IH) and Moderate Hazard (MH).

(in m)	MH	IH	HH
MH	15	30	60
IH		30	60
HH			60

Minimum distances between units are specified by codes, as per the combination of risks. This will ensure, for instance, that flammable vapour released by one unit diffuses to a concentration below the explosive limit when it reaches another unit where a source of ignition exists.

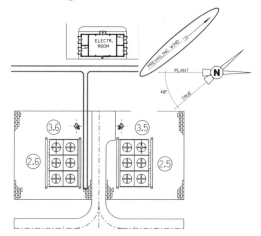

Additionally, units at risk of releasing flammable substances are located away and downwind of equipment that could be source of ignition.

Electrical sub-stations, for instance, will be located upwind of gas coolers. Should a leak occur in the gas coolers, the gas cloud will not reach the electrical sub-station which houses spark generating equipment that could ignite the gas cloud.

Plant layout

Within units, the positions of equipment naturally follow their sequence in the process flow to minimize the length of inter-connections.

Free space is provided around equipment for proper operator access and for maintenance (removal of parts, truck/crane access).

Special attention is given to free space/access for personnel evacuation and fire fighting. Enough space is provided between and around equipment.

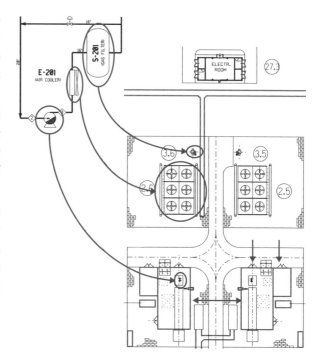

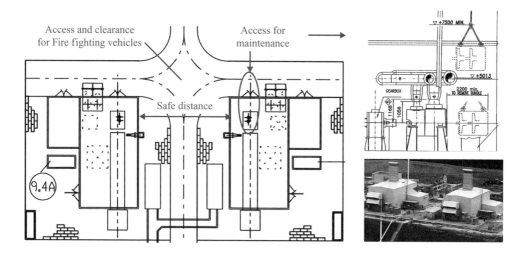

An optimum spacing is found with respect to cost. The size of the plant footprint has indeed a direct impact on the quantities to be purchased and installed: length of pipes, pipe-racks, electrical and instrumentation cables, sewage, fire fighting, roads, paved areas, etc.

Equipment elevations may be dictated by process reasons. A pump, for instance, shall be placed at a lower elevation than the vessel from which it is fed, to ensure proper supply to the pump without cavitation. Another example is that of vessel that is used to collect drains by gravity. It must be located at a lower elevation than the vessel(s) it drains from.

Equipment dimensions will not be available initially. It will become available once the equipment supplier has finalized its design and purchased its sub-equipment. Not only the size of the main equipment should be considered while elaborating the Plot Plan, such as a turbo-compressor unit, but also their auxiliaries, e.g., fuel gas unit, lube oil skid, etc.

Plant layout

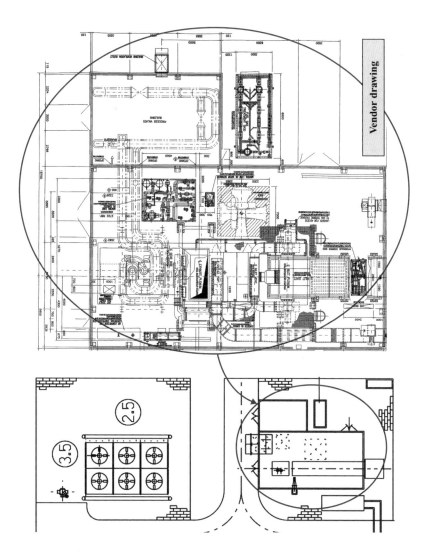

Experience is required to account for all equipment, and estimate their size with accuracy before actual information is available from vendors, in order to be able to freeze the plot plan at an early stage.

Such freeze of the plot plan is essential as it is a pre-requisite for the start of Site activities.

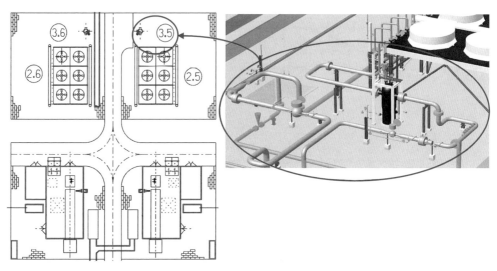

Space for routing of *all* networks (all process and utility pipes, Electrical and Control cables, fire fighting, sewage, pits…) must be duly considered in the Plot Plan. This is the reason why defining a correct layout requires a lot of experience. One indeed needs to have a vision of the entire plant, including all pipes, networks, accesses, etc., before they appear on the drawing boards, thanks to one's experience on previous facilities. All such items will indeed come later on as the design develops and will occupy space that must have been reserved.

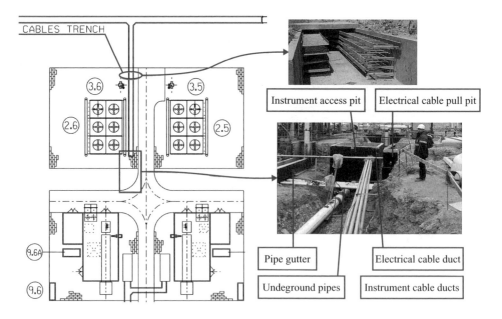

Plant layout

The position of equipment is shown on the **Unit Plot Plan**, by means of coordinates or distance to axis of reference datum point, e.g., inlet nozzle on equipment.

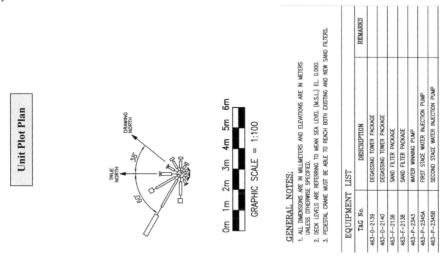

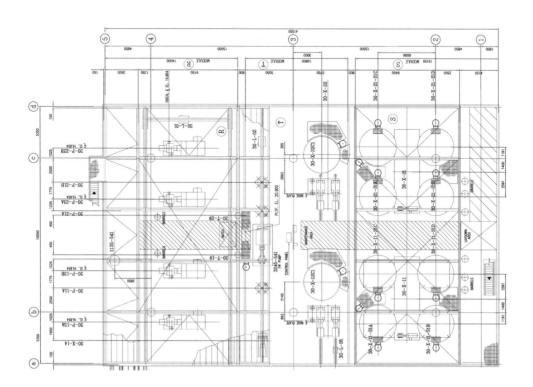

Chapter 6

Health, Safety & Environment (HSE)

Health, Safety and Environment (HSE), also called Loss Prevention Engineering or simply "Safety", works at preventing or minimizing the consequences of accidents linked to the operation of the plant. It is also in charge of ensuring that it complies with legal requirements in terms of release to the environment (gaseous emissions, waste water, noise, solid wastes, etc.).

The first field in which safety is involved is that of the plant process itself. It leads an audit, called a **HAZOP Review** (HAZard and OPerability Review), which is a systematic review of all possible process upsets, verifying that the design incorporates adequate safeguards to allow the process to come back to normal or to safely shut down.

HAZOP reviews are usually carried by Safety with the help of a third party, to avoid conflicts between safety requirements, Contractor and Client's interests. The reviewing team includes Process, Operations, Instrumentation & Control specialists, as many aspects are to be jointly considered.

A systematic method is used for the review: at each point of the process (in each line, in each equipment) the causes and consequences of possible process (pressure/temperature/flow) upsets are identified and evaluated.

The likelihood of an undesirable event happening is evaluated, taking into account the safeguards already included in the design, such as safety automations, alarm, operating instructions.

The consequence that would result from the undesirable event is determined and its severity is ranked using criteria such as the ones shown in the table below.

The risk is then finally evaluated as a combination of the likelihood to happen (frequency) and the severity.

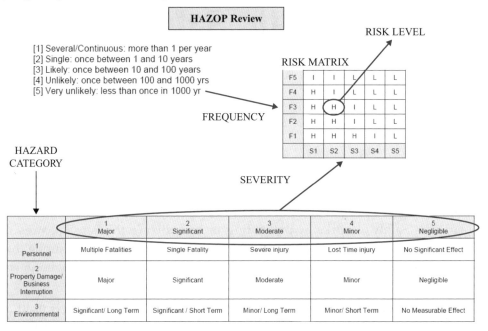

High risks (H) are events with severe consequences and high likelihood to happen. High risks are unacceptable.

The HAZOP team identifies high risks, for which it records that the design must be improved. Precise tracking of the status and expediting of these requested improvements will be made throughout the Engineering phase in order to ensure their implementation.

A typical example of a reviewed item would be the scenario of overflow of a liquid containing vessel.

The team would identify the possible cause (miss operation during filling), consequence (release of product to atmosphere through vessel overflow pipe), existing safeguards (liquid level indicator, operating instructions, high level alarm).

Health, Safety & Environment (HSE)

The Frequency of occurrence will be estimated considering likelihood of error by operator, malfunction of the level sensor, etc.

The Severity will depend on the type of product released to atmosphere (personnel and/or environmental hazard).

Should Frequency * Severity (risk) be found high, the HAZOP team would prescribe an action from the design team (a check to be done, a calculation to be made, a change to the design such as the addition of a safeguard, etc.).

The form below shows a typical HAZOP worksheet, issued after the review as part of the **HAZOP report**. The first item does not require any action by the design team. The second item requires an action (addition of a non-return valve), which Engineering has implemented as shown on the P&ID.

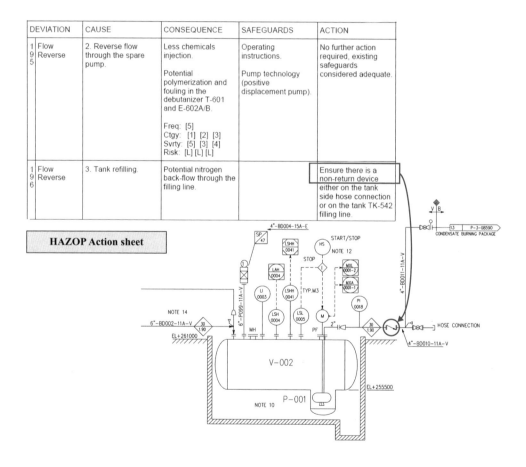

HAZOP Action sheet

Early incorporation of additional requirements resulting from the HAZOP is essential to minimize the amount of design reworks they generate.

Second only to process safety is the safe layout of the facility. Explosion and fire hazards exist in Oil & Gas facilities due to the flammable and explosive inventories handled. Adequate design considerations, in particular in the field of layout (relative positions of equipment) and spacing (minimum distances between equipment), can reduce the risk or consequence of such events.

Explosion and fire damage can indeed be significantly reduced with proper layout as explosion overpressure and fire radiation intensity rapidly decrease with distance. Minimum distances are specified between units and equipment based on the risk levels.

Safety will also review the layout of the plant to ensure sufficient space is provided for escape of personnel in case of emergency and for access for fire fighting.

Please refer to the Plant Layout section for details of safety considerations in plant layout.

The Fire Fighting system of the plant is designed by Safety. Such system comprises both passive and active fire fighting means.

Active fire fighting system consist of the fire water system, a pressurized water ring feeding hydrants, fire monitors (for manual fire fighting) and the deluge system (for automatic fire fighting).

The deluge system consists of spray nozzles (sprinklers) arranged around the equipment, that will automatically spray water on the equipment upon detection of fire. The detection itself is done by fusible plugs located around the equipment, that melt when subject to heat.

The purpose of the water spray is not to extinguish the fire but to cool down the equipment, for instance a pressure vessel, to prevent the steel from loosing its strength at elevated temperature which could lead to the collapse of the vessel and loss of containment.

The quantity of fire fighting water is determined in the **fire water demand calculation note**. The plant area is first divided into fire zones.

Health, Safety & Environment (HSE)

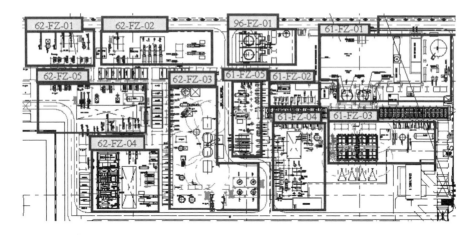

The water demand calculation is then calculated on the basis of a fire in one of the fire zone, with all fire fighting equipment in operation in this fire zone.

The deluge water demand is calculated from the number of sprinkler nozzles, itself a function of the the surface areas of the protected vessels.

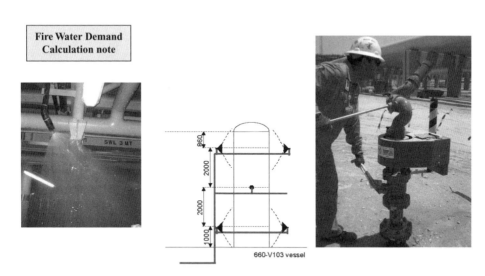

Fire Water Demand Calculation note

Item	Calculated flow rate	Selected flow rate
Maximum flowrate for deluge system	1117 m^3/h	745 m^3/h
Flowrate for monitors (6)	684 m^3/h	456 m^3/h
Flowrate for hoses (4)	228 m^3/h	228 m^3/h
Common facilities area (Unit 660) total firewater demand		1429 m^3/h

Health, Safety & Environment (HSE)

The fire water system is depicted by the Safety on the **Fire Water Piping & Instrumentation Diagrams (P&IDs)**.

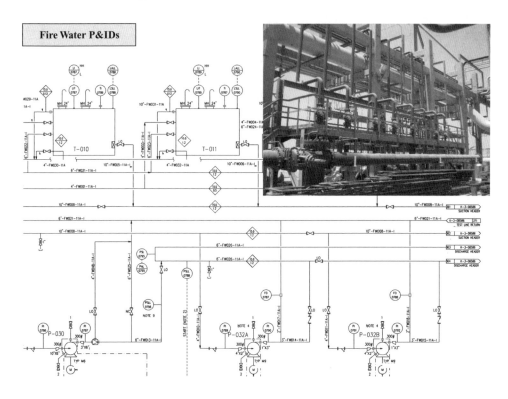

Arrangement of deluge nozzles around equipment is shown on the **Deluge system arrangement drawings**.

Health, Safety & Environment (HSE)

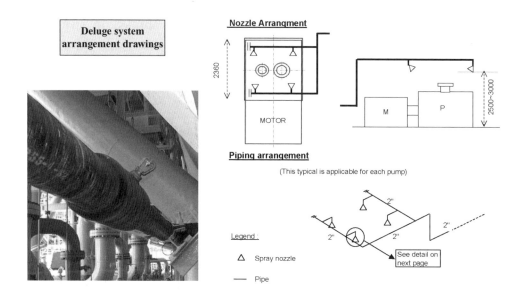

Deluge system arrangement drawings

The location of the fire fighting equipment is shown on the **Fire fighting equipment location drawings**.

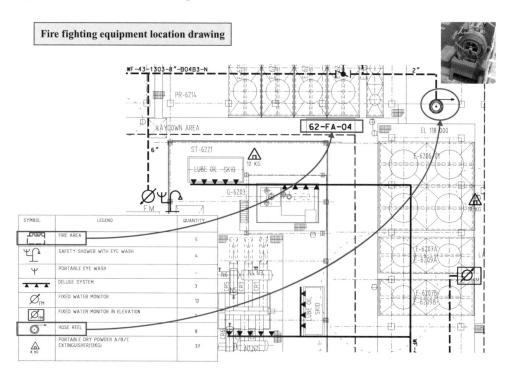

Passive fire fighting, by means of fireproofing, is applied to structures supporting equipment and pipes. Protection of such structures will prevent/delay the fall of critical equipment or pipes therefore avoiding the escalation of the incident.

In order to define which structures shall be fireproofed, Satey proceeds as follows:

It first establishes the list of equipment generating a fire hazard, such as equipment containing a significant volume of flammable liquid, etc.

Each such equipment creates a "fire scenario envelope" in its surroundings. The various envelopes are consolidated and structures located inside the overall envelope are identified.

Not all structures within the envelope shall be fireproofed, but only the ones supporting equipment and pipe whose collapse could lead to incident escalation or large damage. This would include for instance a large and heavy tank, even if merely containing water, located at height.

Extent of fireproofing of the structures is defined by means of typical drawings such as the ones shown here.

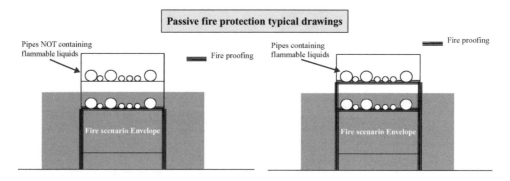

Fire proofing can be done by applying a special coating, or concrete, in which case requirements are addressed to the civil engineer who develops the required standard drawings.

Health, Safety & Environment (HSE)

Fire proofing

Fire fighting includes a Fire and Gas detection system, which activates alarms and performs automatic actions, such as electrical isolation, in case of fire and gas detection.

Health, Safety & Environment (HSE)

Safety defines the number, location and type of Fire and Gas detectors both in process areas and inside buildings and shows the same on the **Fire & Gas detection layout drawings**.

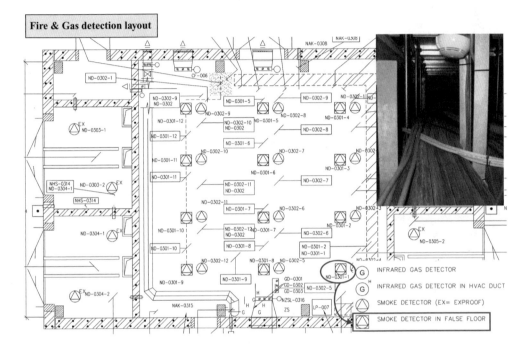

Fire & Gas detection layout

Safety defines the emergency actions, such as process shutdown, electrical isolation, etc. upon fire or gas detection. These actions and their initiators are shown on the **ESD logic diagrams**.

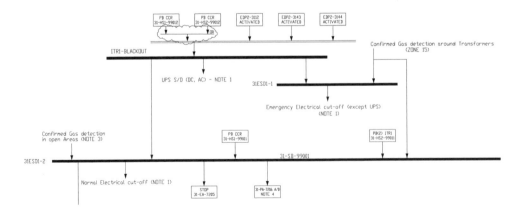

Health, Safety & Environment (HSE)

The detailed logic is shown on the **Fire and Gas Matrix**.

CAUSES				EFFECTS							
Location	Causes	Voting	Setpoint	Local F&G panel	Master F&G panel	Gate house F&G panel	Fire building F&G panel	Audible and visible Fire alarm	Audible and visible Gas alarm	SD-2	Close electrical substation 27.1 fire dampers and stop HVAC
Compressor unit 27.1 - Electrical substation											
Transformers	Optical smoke detector	1 out of 2		X	X	X	X				
		2 out of 2		X	X	X	X	X			
	Manual Fire Alarm Station	1 out of 1		X	X	X	X	X			
HVAC inlet	Infrared gas detection	1 out of 3	10% LFL	X	X	X	X				
		1 out of 3	20% LFL	X	X	X	X				
		2 out of 3	10% LFL	X	X	X	X		X		
		2 out of 3	20% LFL	X	X	X	X		X		X
Electrical room and false floor	Optical smoke detector	1 out of 2		X	X	X	X				
		2 out of 2		X	X	X	X	X			X
	Manual Fire Alarm Station	1 out of 1		X	X	X	X	X			

In the example shown above, detection of gas in the air inlet of the building ventilation system will cause the ventilation fan to stop and the damper (shutter of the ventilation duct) to close. Indeed, the equipment located inside buildings is not designed to work in an explosive atmosphere.

Safety identifies the areas of the plant where explosive atmospheres could form. This is based on the identification of known sources of release and potential sources of leaks.

Sources of release include storage tanks, vents of equipment and instruments, etc. Potential sources of leaks include flanged connections in pipework etc. The extent of the explosive atmosphere around the source is assumed using a set of rules, for instance a radius of 3 meters around an instrument vent, etc.

Hazardous area classification drawings are prepared showing areas where an explosive atmosphere could be present, along with the likeliness of presence (Zone 0/1/2).

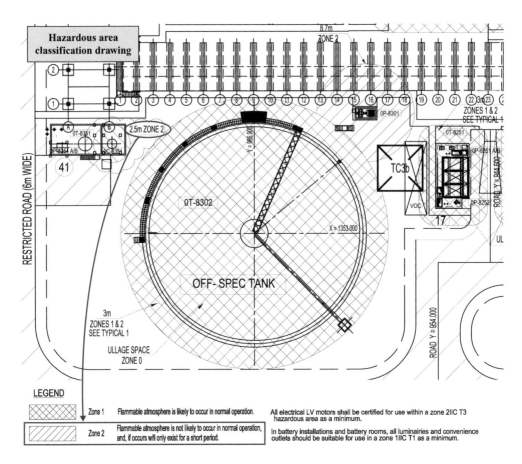

Electrical equipment located in hazardous areas must be of a special design so that they are not a source of ignition. Such special design provides various degree of protection against the risk of being a source of ignition.

The required degree of protection is determined based on the classification (zone 0 > 1 > 2) of the area where the equipment is located.

Protection could be achieved by different designs such as:

- explosion proof, referred to as "d": the equipment is enclosed inside a heavy duty enclosure that would contain an explosion and avoid its propagation,
- increased safety, referred to as "e": the equipment is designed not to generate any spark,
- intrinsic safety, referred to as "i": the amount of energy created by a spark in the equipment is not high enough to ignite the explosive atmosphere,
- etc.

Besides this level of explosion protection, Safety specifies the composition of the explosive atmosphere to which the equipment could be exposed. The nature of the explosive atmosphere has indeed a direct impact on the minimum ignition energy. An atmosphere of hydrogen, such as the one that could develop in a battery room during charging, requires much less energy to ignite than a natural gas atmosphere for instance. The nature of the atmosphere is specified by reference to a gas group, e.g., IIC for hydrogen, etc.

Finally, Safety specifies the maximum temperature authorized on the equipment surface. Indeed, the explosive atmosphere will ignite if it comes in contact with a temperature above its self-ignition temperature. This again depends on the composition of the explosive atmosphere: methane self-ignition temperature is around 600°C whereas that of ethylene is 425°C.

The maximum equipment surface temperature is specified by means of a temperature class, e.g., T3 means maximum surface temperature of 200°C.

Electrical equipment protected against explosion is clearly marked by means of an international code encompassing the information above:

Explosion proof

Increased Safety

The **Quantitative Risk Analysis (QRA)** is a systematic way to assess the hazardous situations associated with the operation of the plant. The analysis is related to release of hazardous materials to atmosphere that can cause damage to people or equipment, e.g., due to explosion, fire, etc.

Each accidental event is plotted inside a risk matrix, according to its frequency and severity.

Health, Safety & Environment (HSE)

Action is required for any event falling in the "Intolerable Risk Area" of the matrix. Its frequency or consequences must be reduced to bring it into the "ALARP (As Low As Reasonably Practicable)" or "Acceptable" risk areas, through risk reduction measures.

The first step of the QRA is to perform a hazard identification.

In the example that follows, the hazard reviewed is that of an explosion due to leak from piping. The cause could be material detects, construction errors, corrosion, maintenance overlook, etc.

The plant is divided into individual isolable sections of similar hazardous material, process conditions and location. The section considered here is the building housing a compressor.

The inventory of each component from which the leak could originate (pipes, flanges, pumps, valves, instruments...) is made. Frequency of leak of individual components is taken from statistical data found in the literature, for various leak size, e.g., 5% of component bore size, etc.

The sum of the individual component leak frequencies and sizes give the overall plant section leak frequency and size.

Quantitative Risk Analysis (QRA)

Case study: Gas leak from random piping component rupture

Cause: installation error, corrosion, material defect...

Possible consequence: Dispersion without ignition / jet fire / flash fire / explosion

Section considered: Compressor building

Σ risk components
*failure rate
(from statistics)

Step 1:

Identification and characterisation of initiating events

Gas leak inside compressor buidling due to component rupture	Hole size (% of component section)		
	5%	20%	Full
Frequency (event/year)	1,11E -01	5,06E -04	6,83E -05
Outflow rate (kg/s)	5,7	90,8	2270,0

Health, Safety & Environment (HSE)

Release of hazardous material to atmosphere can give rise to different effects, such as simple dispersion without harm or on the contrary fire, explosion, etc. This depends on a number of factors, such the presence of ignition sources, the degree of confinement, etc. It is the purpose of the second step of the QRA to evaluate the probability of each possible consequence.

The various scenarios are shown on event trees. The frequency of each event is factored by the probability of the subsequent one, resulting in the frequency of the various possible ultimate consequences.

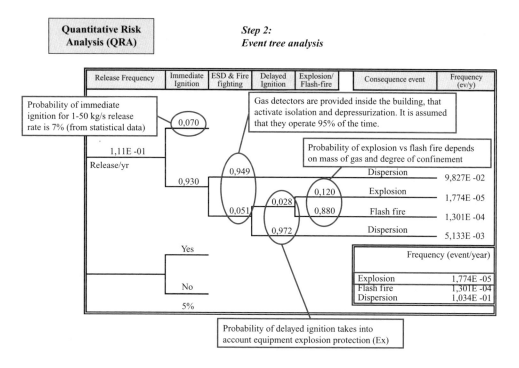

The third step of the QRA is to evaluate the effects of each accidental scenario. Consequences are expressed in terms of reference values of overpressure, heat radiation, etc.

Health, Safety & Environment (HSE)

Quantitative Risk Analysis (QRA)

Step 3:
Consequence evalutation

Overpresssure (bar)	0.2	0.1	0.01
Distance (m)	96	167	1270

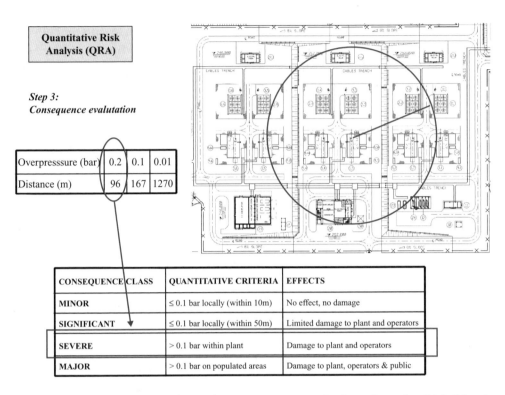

CONSEQUENCE CLASS	QUANTITATIVE CRITERIA	EFFECTS
MINOR	≤ 0.1 bar locally (within 10m)	No effect, no damage
SIGNIFICANT	≤ 0.1 bar locally (within 50m)	Limited damage to plant and operators
SEVERE	> 0.1 bar within plant	Damage to plant and operators
MAJOR	> 0.1 bar on populated areas	Damage to plant, operators & public

The risk is ranked in a class of consequences and plotted on the Risk Matrix to check its acceptability.

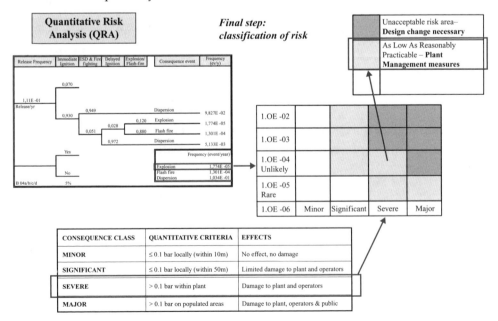

CONSEQUENCE CLASS	QUANTITATIVE CRITERIA	EFFECTS
MINOR	≤ 0.1 bar locally (within 10m)	No effect, no damage
SIGNIFICANT	≤ 0.1 bar locally (within 50m)	Limited damage to plant and operators
SEVERE	> 0.1 bar within plant	Damage to plant and operators
MAJOR	> 0.1 bar on populated areas	Damage to plant, operators & public

Health, Safety & Environment (HSE)

The Quantitative Risk Assessment results in requirements, such as blast resistance of buildings, reinforcement of structures supporting safety critical elements, etc., which are incorporated in the design.

The impact of the plant on the environment is specified and evaluated by the HSE discipline.

An **ENVID** (ENVironmental aspects IDentification) review is performed to identify all environmental aspects of the plant, i.e., all equipment having a potential impact on the environment.

Aspect	Health	Air	Water		Raw material	Waste
		Gaseous emissions	Resource Consumption	Liquid effluents	Petroleum/gas /Chemicals	
Relief (flare/vent)	Noise*	CO, NO_x, PM, SO_2, VOC				
Power generation		CO, NO_x, PM, SO_2			Fuelgas	
Gas compression	Noise*	Fugitive VOC			Gas	
Fresh water	Potable		X			
Cooling water	Legionella		X	Effluent Water Temperature	Biocides, pH Control	
Effluent water (open drains/ treatment Plant)				Hydrocarbons, Suspended Solids		Biosludges, Oily sludge

The review covers, for each aspect, the corresponding environmental concerns (noise, NOX emission, energy consumption, waste generation...) and the measures that are implemented in the design to control the environmental impact.

The **Health and Environment Requirements** specification states the requirements for each of the identified environmental aspect: regulatory standards, limits for all emissions (contaminants in discharged water, pollutants in gaseous discharges, etc.), design dispositions to limit/monitor pollutants for each type of emission/effluent discharge, ambient air quality, noise limits, disposition for disposal of hazardous wastes, etc.

Effluent Quality Criteria for Discharge into Sea Organic Species				
Parameter	Symbol	Units	Monthly Average	Maximum Allowable
Oil & Grease		mg/l	5	10
Phenols		mg/l	0.1	0.5
Total Organic Carbon	TOC	mg/l	50	75
Halogenated hydrocarbons and Pesticides		mg/l	***	

The above requirements are fed back into the design (water segregation and treatment system, height of exhaust stacks) and addressed to equipment vendors (limits of NOx for gas turbines, etc.).

Later in the Project, an **Environmental Impact Assessment** is performed to verify that the design complies with the above requirements.

It includes an analysis of the dispersion of atmospheric pollutants released by the plant to evaluate the impact of the plant on the surrounding air quality. It entails an inventory of all sources of atmospheric emissions (machinery exhausts, etc.), and the modelling of the atmospheric dispersion according to local meteorological data. It results in the calculation of the levels of ground concentration of atmospheric pollutants at various distances from the plant, e.g., within the facility, in nearby populated areas, etc.

Health, Safety & Environment (HSE)

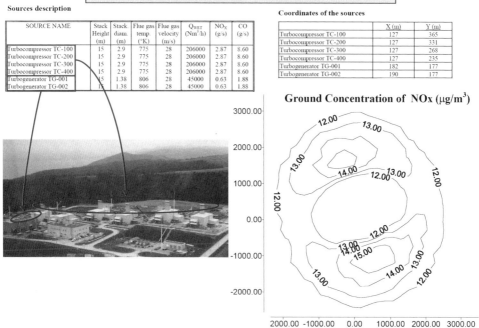

Environmental Impact Assessment (Air quality dispersion analysis)

Sources description

SOURCE NAME	Stack Height (m)	Stack diam. (m)	Flue gas temp. (°K)	Flue gas velocity (m/s)	Q_{WET} (Nm³/h)	NO_x (g/s)	CO (g/s)
Turbocompressor TC-100	15	2.9	775	28	206000	2.87	8.60
Turbocompressor TC-200	15	2.9	775	28	206000	2.87	8.60
Turbocompressor TC-300	15	2.9	775	28	206000	2.87	8.60
Turbocompressor TC-400	15	2.9	775	28	206000	2.87	8.60
Turbogenerator TG-001	15	1.38	806	28	45000	0.63	1.88
Turbogenerator TG-002	15	1.38	806	28	45000	0.63	1.88

Coordinates of the sources

	X (m)	Y (m)
Turbocompressor TC-100	127	365
Turbocompressor TC-200	127	331
Turbocompressor TC-300	127	268
Turbocompressor TC-400	127	235
Turbogenerator TG-001	182	177
Turbogenerator TG-002	190	177

The scope of the Environmental Impact Assessment covers emissions in normal operation only. Accidental emissions and their impact on the facilities or populations is out of the scope and is covered in the Quantitative Risk Assessment.

The environmental impact assessment also includes a **Noise study**. It starts with the inventory of all noise sources. Noise levels are then obtained from reference data base during preliminary studies, then from each equipment vendor after purchase. A computer is used to run a model of the noise dispersion. Both noise sources and barrier elements with noise screen effect such as buildings, are entered in the model. The noise level at each location of the plant is evaluated. Verification is done that noise levels in working areas, and at the facility's boundaries, are within the safe/legal limits.

The noise study records the bases and results of noise calculations. Equipment noise insulation requirements are derived from the noise study. The results of the noise study are shown on the **Noise map**.

Noise Map

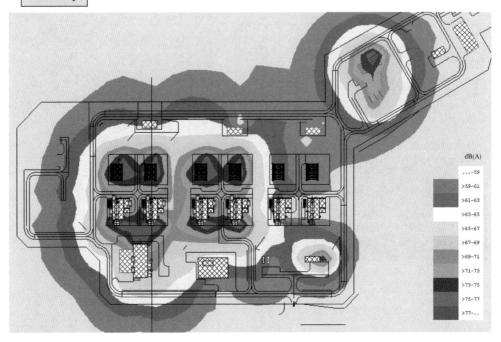

Finally, the Environmental Impact Assessment includes a waste management study. The wastes generated by the plant are inventoried and the possible options for recycling, treatment or disposal are studied based on existing local waste recycling/treatment/disposal facilities. This study allows to size the temporary waste storage area required on site.

Chapter 7

Civil engineering

The first step of civil engineering for an On-Shore plant is to know the Site and the type of soil on which the plant will be built. A survey is required to collect topographical, hydrological, geological and geotechnical data. A **Soil Investigations Specification** is prepared by the Geotechnical Engineer to define the scope of this survey. It will include soil investigations, by means of geotechnical and geophysical methods, to collect a good understanding of the type of soil and its variability over the plant area. The type of soil determines the type of equipment required for excavations (excavators/explosives) and the type of foundations (shallow/deep) required for the plant equipment.

It will also include the identification of any local geo hazards, such as seismic hazard, collapsible soil, underground cavities, etc. and the definition of the soil geotechnical parameters to be used for the sizing of the foundations, such as soil bearing capacity.

Specific studies are performed underneath critical structures, such as large storage tanks, tall columns, etc.

The geotechnical engineer will also provide specific recommendations to the engineer with respect to, for instance, reinforcement of slopes to be done in backfilled areas, etc.

Civil engineering

Once the overall layout (General Plot Plan) of the facility is defined, the first Site activities can start. These are the earthworks, which consist of levelling the site up to the required elevation.

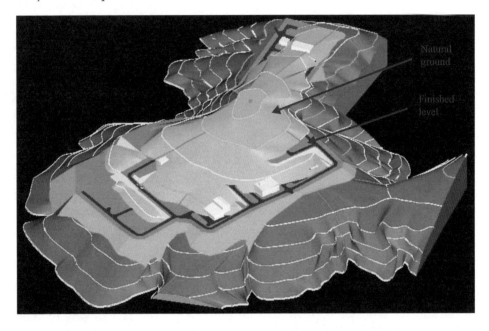

Earthworks drawings are produced, such as the **Grading Plan**, showing the natural ground elevation and the final desired elevation.

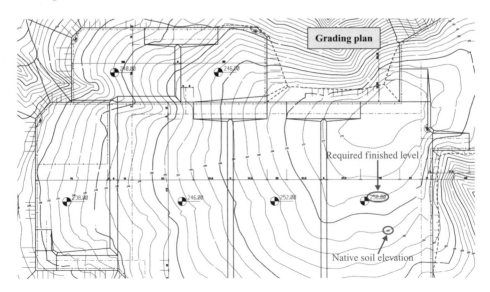

Earthworks equipment will then excavate/fill in order to reach the required finished level.

Once the Site is levelled, local excavation can be done and foundations of main equipment can be cast. Indeed, the main equipment foundations are the deepest undergrounds to be installed hence they have to be installed first.

Design of equipment foundation requires Vendor information such as footprint, location of anchor bolts, static and dynamic loads, etc. The vendor determines these loads, which are the basis for the sizing of the foundation.

Civil engineering

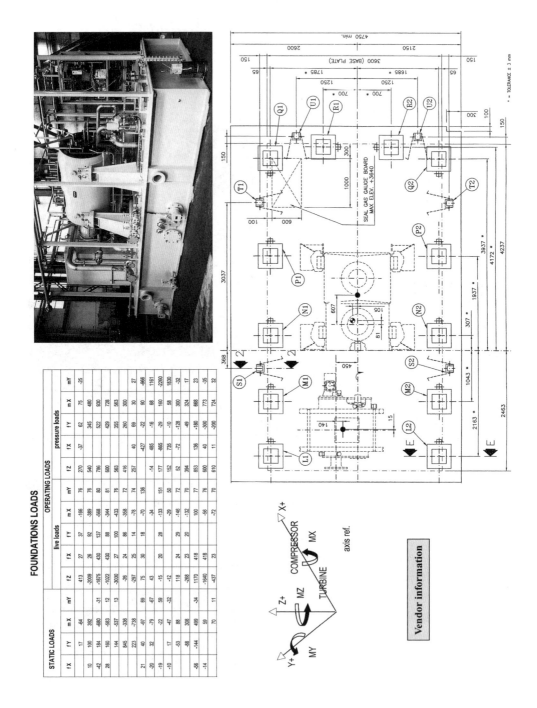

Civil engineering 75

The civil engineer designs the foundation using a computer software.

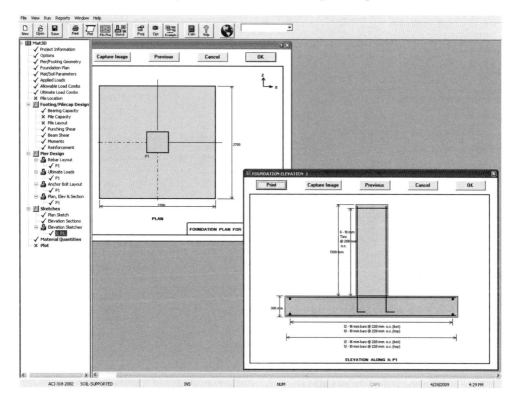

The type (piles, etc.) and size of the foundation depend on soil characteristics (bearing capacity, etc.). The bases of design and results of calculations are recorded in the **foundation calculation note**.

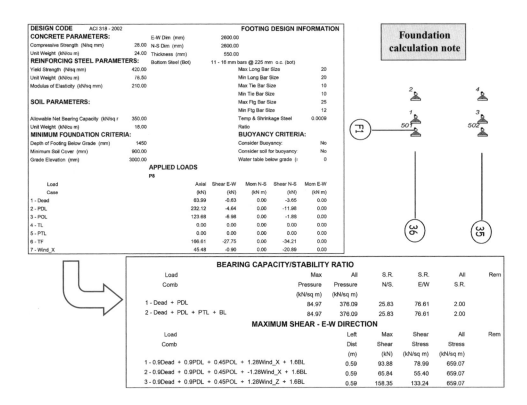

From the above design results the size of the foundation, its shape, dimensions, depth, and amount of re-inforcement.

Foundation drawings are produced, which show the dimensions and depth of the foundation, the position, number and size of re-inforcing bars, and the position of anchor bolts to be cast in the foundation.

Reinforcement drawings and **Formwork drawings** are usually issued as separate drawings.

Civil engineering

Foundation drawing

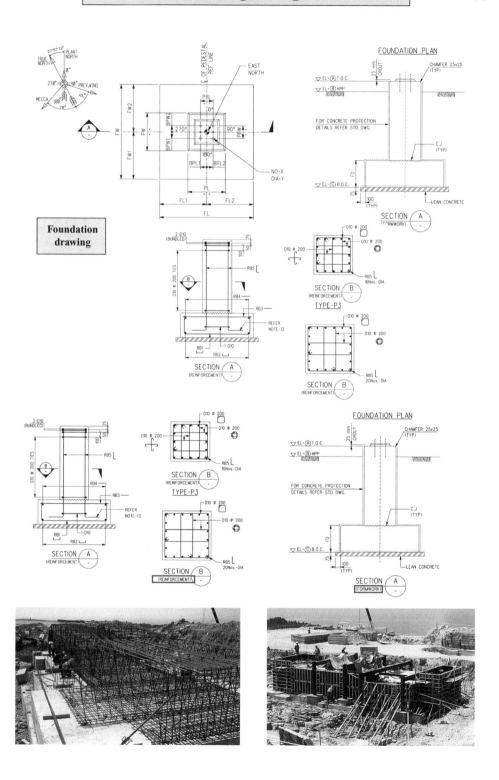

Besides drawings, Civil issues **Civil works specifications**, for each trade, e.g., site preparation, concrete works, roads, buildings, etc. which define the materials to be used, how the work shall be done, the inspections and testing requirements, etc.

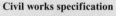

Pre-fabrication is done to the maximum possible extent in order to reduce installation time. Concrete indeed requires around 2 weeks to dry before it can be buried. For the case of a foundation cast in-situ for instance, the excavation, which occupies a large footprint, needs to remain open for such period of time, which prevents surface works to proceed. Pre-fabrication of the foundation would avoid that and allow immediate backfill after installation.

Besides specific concrete constructions which are one-off and customized to a particular equipment, civil also produces generic concrete items. **Civil standard drawings**, such as the one showing standardized pipe support foundations here, are issued for that purpose. Such standardisation allows mass production at the pre-fabrication yard.

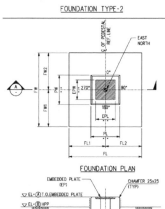

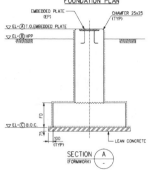

Civil engineering

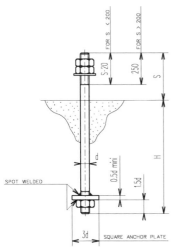

Civil also issues **Construction standards**, such as the one shown here for anchor bolts, which show repetitive arrangements.

Similarly to foundations for an On-Shore facility, deck structural drawings are produced for an Off-Shore facility.

The deck structure is made of the primary structure, which comprises the main girders making the deck frames, and the connection between the decks (legs), the secondary structure, made of beams supporting equipment, and tertiary structure, made of small beams supporting plating.

Layout studies determine the number, size and elevation of deck levels and the main equipment location. Together with equipment weights, it allows the Structure discipline to perform its design, calculations and to issue the **Primary Steel Structure drawings**.

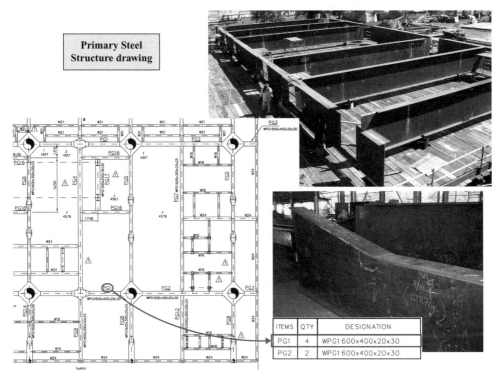

Primary Steel Structure drawing

ITEMS	QTY	DESIGNATION
PG1	4	WPG1600x400x20x30
PG2	2	WPG1600x400x20x30

The primary structure (welded plate girders forming the deck frame, deck legs, etc.) is made out of steel plates that are a long lead item. Indeed, such steel has special properties (high strength, through thickness properties), requires special tests and must come from a mill that has been duly qualified.

The primary steel structure material take-off is therefore issued early in the project to quantity all necessary steel plates.

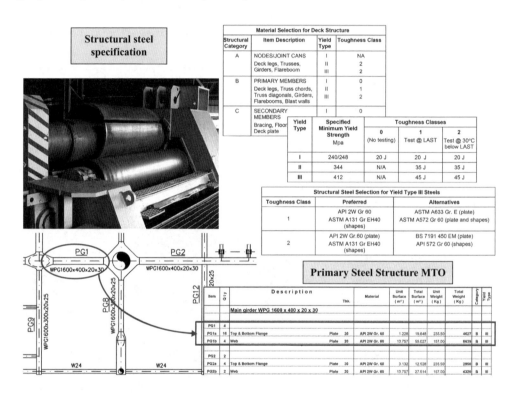

Secondary structure drawings are issued next, which show the main equipment support beams, and the associated bill of material, which has of a shorter lead time than primary steel.

Civil engineering

Secondary Steel Structure drawing

			MATERIAL LIST						
ITEMS	QTY	DESIGNATION	MATERIAL	L.A.V. UNIT	L.A.V. TOTAL	WEIGHT UNIT	WEIGHT TOTAL	CAT	TYP
B1	4	W24x146	ASTM A131DH/EH36	4.980	19.920	217	4323	B	II
B2	1	W24x146	ASTM A131DH/EH36	4.734	4.734	217	1027	B	II
B60	1	W24x146	ASTM A131DH/EH36	3.564	3.564	217	773	B	II

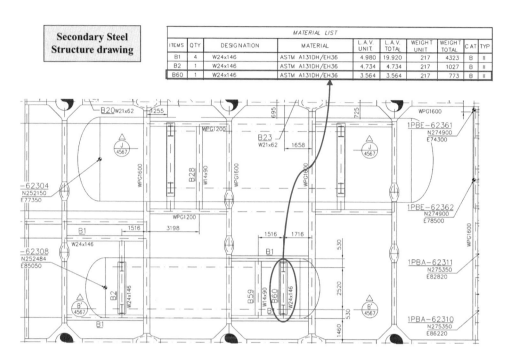

In On-Shore facilities, besides equipment supporting structures, long stretches of large steel structures supporting pipes, called pipe-racks, are found.

Requirements for these structures (location, width, number and elevation of levels, etc.) and input for their design (number of pipes to be supported, weight, operating loads) are defined by Piping.

Good communication between Piping and Civil is essential to optimize their design and include contingencies in order to avoid changes at a later stage, when piping studies will have progressed.

Large piping operating loads, such as loads at piping fixed points, thermal loads from low or high temperature lines (subject to high expansion), etc. are calculated by Piping Stress analysis group and advised to Civil. Other piping loads are estimated by Civil.

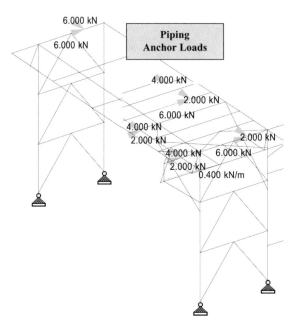

To these piping loads are added external loads such as seismic and wind loads.

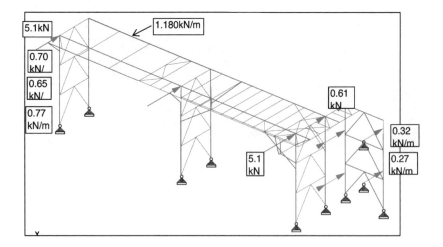

The structure is designed using a 3D modelling and calculation software.

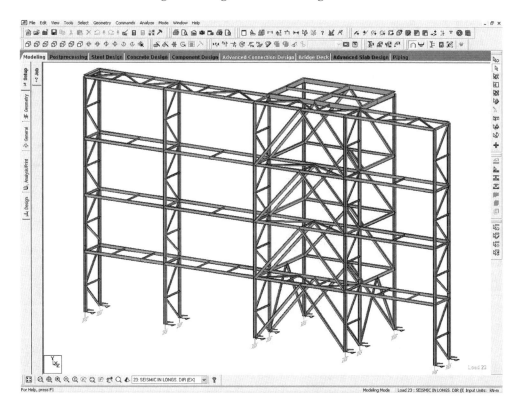

Design of the structure includes sizing of main members, selection of connection type (pin/moment) between the members, provision of secondary members, such bracings for stability, etc. It results in a **structural calculation note**, which shows the design input and results (stress ratios in structure members, deflection of members).

Civil engineering

Calculation note

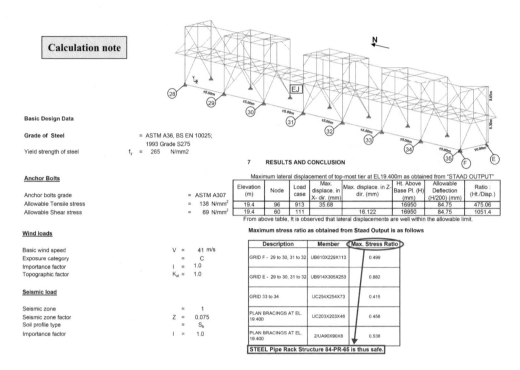

Basic Design Data

Grade of Steel = ASTM A36, BS EN 10025; 1993 Grade S275
Yield strength of steel f_y = 265 N/mm²

Anchor Bolts

Anchor bolts grade = ASTM A307
Allowable Tensile stress = 138 N/mm²
Allowable Shear stress = 69 N/mm²

Wind loads

Basic wind speed V = 41 m/s
Exposure category = C
Importance factor I = 1.0
Topographic factor K_{zt} = 1.0

Seismic load

Seismic zone = 1
Seismic zone factor Z = 0.075
Soil profile type = S_b
Importance factor I = 1.0

7 RESULTS AND CONCLUSION

Maximum lateral displacement of top-most tier at EL19.400m as obtained from "STAAD OUTPUT"

Elevation (m)	Node	Load case	Max. displace. in X- dir. (mm)	Max. displace. in Z- dir. (mm)	Ht. Above Base Pl. (H) (mm)	Allowable Deflection (H/200) (mm)	Ratio : (Ht./Disp.)
19.4	96	913	35.68		16950	84.75	475.06
19.4	60	111		16.122	16950	84.75	1051.4

From above table, It is observed that lateral displacements are well within the allowable limit.

Maximum stress ratio as obtained from Staad Output is as follows

Description	Member	Max. Stress Ratio
GRID F - 29 to 30, 31 to 32	UB610X229X113	0.499
GRID E - 29 to 30, 31 to 32	UB914X305X253	0.882
GRID 33 to 34	UC254X254X73	0.415
PLAN BRACINGS AT EL. 19.400	UC203X203X46	0.458
PLAN BRACINGS AT EL. 19.400	2/UA90X90X8	0.538

STEEL Pipe Rack Structure 84-PR-65 is thus safe.

The civil designer then prepares the **steel structure design drawings**, which are issued to the manufacturer of steel.

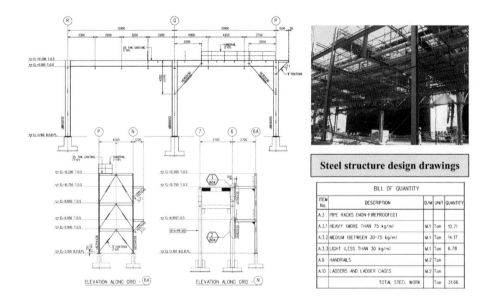

Steel structure design drawings

ITEM No.	DESCRIPTION	D/M	UNIT	QUANTITY
A.3	PIPE RACKS (NON-FIREPROOFED)			
A.3.1	HEAVY (MORE THAN 75 kg/m)	M.1	Ton	10.71
A.3.2	MEDIUM (BETWEEN 30-75 kg/m)	M.1	Ton	14.17
A.3.3	LIGHT (LESS THAN 30 kg/m)	M.1	Ton	6.78
A.9	HANDRAILS	M.2	Ton	
A.10	LADDERS AND LADDER CAGES	M.2	Ton	
	TOTAL STEEL WORK		Ton	31.66

The steel structure manufacturer models the structure in all details, including all connections between steel members, and issues shop drawings, such as the one shown below, to its fabrication shop.

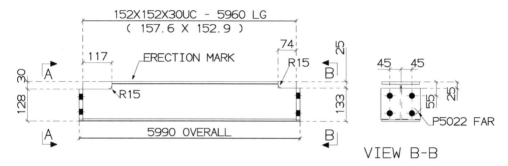

One shop drawing is produced for each structural member, showing all fabrication details, such as exact dimensions, position of gussets, positions and number of holes for bolts, etc. There is also usually a direct transfer of all fabrication data from the design office 3D model software to the numerical control fabrication machinery.

The manufacturer issues the **Erection drawings**, which show the overall view of the structure, together with the arrangement of the various steel members, identified by their piece marks.

Civil engineering

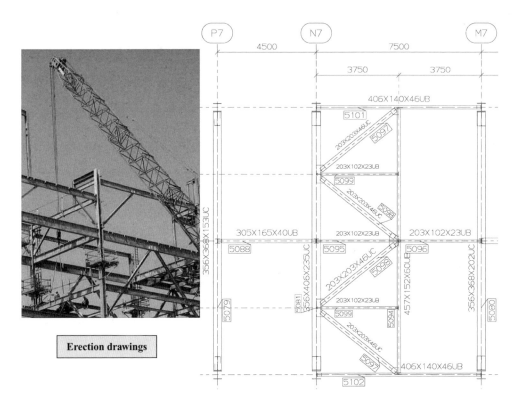

Erection drawings

Identification is key. A given steel structure may come in as many as one thousand pieces, reaching the erection Site by several different truck loads, spread during storage before erection in very extended lay down areas.

On top of equipment and piping supporting structures, civil provides small platforms for operator access (to equipment, instrument, valves, etc.).

Corresponding access requirements (location, elevation, dimensions of operating stages) are identified and defined by Piping installation discipline. Civil discipline implements these requirements by designing the corresponding small structures. A standard design is produced, which will be applied to all these repetitive items, and to the associated handrails, stairs, ladders, etc.

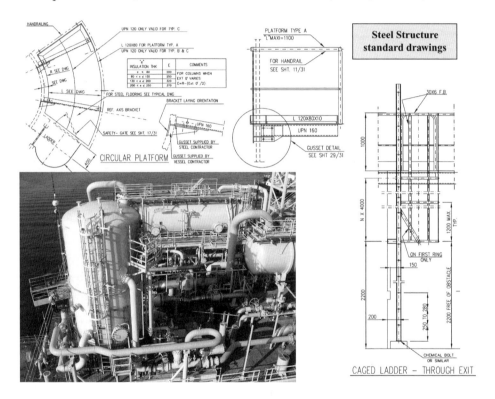

Civil Works Installation (CWI) drawings show the layout of all underground objects and networks. These are very detailed drawings showing, for each area, the location and elevation of the numerous underground objects: foundations (of equipment, structures, buildings, pipe supports, etc.), networks (process and utility services, cables, sewage, pits, roads, etc.).

Production of the CWI drawings ensures coordination of all underground objects in order to anticipate and prevent interferences.

Civil engineering

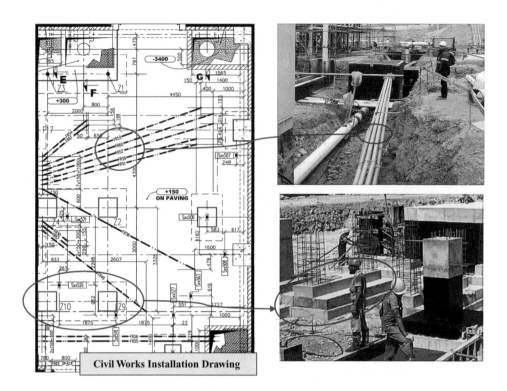

Civil Works Installation Drawing

Priorities exist among the various underground objects and networks. The civil engineer locates the priority ones first, the next priorities will be located in the remaining available space. The sequence is as follows:

- Main equipment and pipe-racks foundations come first, as the main equipment positions are determined by the facility layout and cannot be changed,
- Gravity underground piping, such as sewage, comes second, as it must be sloped hence there is no flexibility in its routing. The space occupied by access pits, provided for cleaning at every change of direction shall also be accounted for,
- Underground pressure piping comes next, as its length must be minimized to reduce costs,
- Then come cables. Width and routes of cable trenches are advised by the corresponding disciplines (Electrical/Instrumentation/Telecom). The space occupied by cable pulling pits, duct banks at road crossing is also considered.

Civil engineering

Civil Works Installation drawings are issued multiple times, to allow Site works to proceed step-wise. As the design progress, underground objects and networks are progressively designed, positioned and shown on the CWI drawings. Site installation starts by the deepest underground items. Accordingly, the CWI are first issued with the main equipment foundation only, then with added rain water collection and underground piping networks, then with cable trenches, pipe supports, etc. and lastly with paving.

Before Civil can issue the paving drawings, all undergrounds must be defined and included. Indeed, after paving is cast, no underground can be installed. This requires the civil engineer to collect information from all disciplines. Civil will, for instance, collect information about pipe support location and loads and incorporate the corresponding foundations or re-inforcements in the paving drawings.

Building design also falls within the scope of the civil engineer. It starts by the architectural definition of the building, i.e., size/number of rooms, etc. which comes from the building function.

An Electrical sub-station, for instance, will be sized according to the number and size of housed switchboards, including provision for future ones. It will also be specified a false floor for cable routing, wall and floor openings for cable penetrations, etc.

The architectural requirements are grouped and shown on the Architectural drawings.

Civil engineering

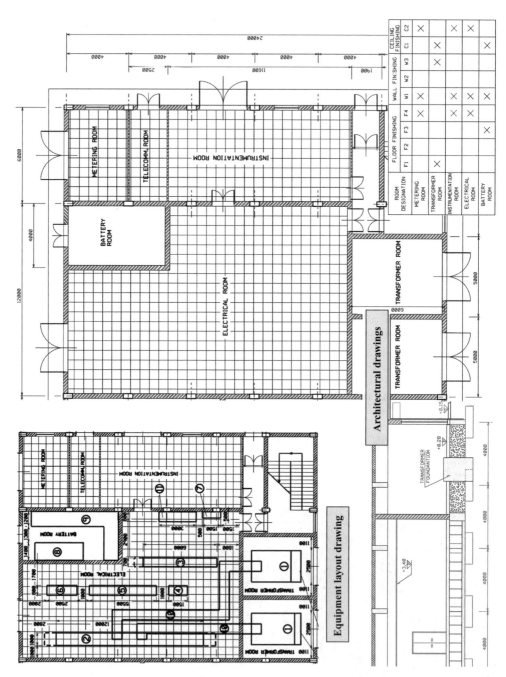

The building detailed design follows from the architectural requirements. Drawings in all trades, with a very high level of details, are produced.

Production of **building detail drawings**, such as the layout of Telecom equipment and cables, is usually sub-contracted to the building construction contractor.

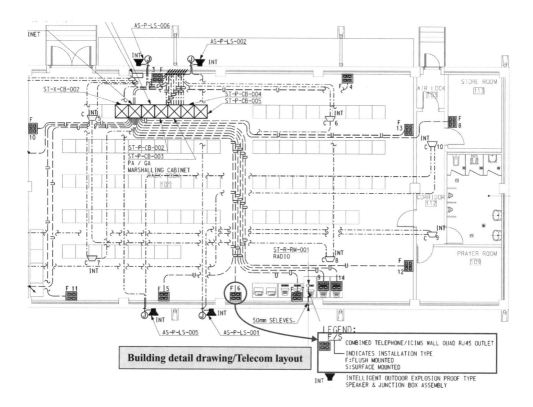

Building detail drawing/Telecom layout

The Heating, Ventilation and Air Conditioning (HVAC) system is also part of the building design and in the Civil engineer's scope.

The HVAC system of a building is designed to provide the required climate inside the building/room.

Examples of climate control requirements are:
- forced ventilation, for mechanical equipment generating heat,
- heating (winter) & air-conditioning (summer) for rooms where personnel could be present,
- ventilation (heat evacuation) and air-conditioning (humidity control) in electrical equipment rooms,
- overpressure maintenance in electrical sub-stations located inside process units (to prevent dust/flammable gas to enter the building).

These requirements are further refined, for each room of each building, depending on its function:

- temperature to be maintained in permanently vs temporarily manned rooms (control room, offices vs corridors, change rooms, etc.),
- temperature to be maintained in equipment rooms.

The design of the HVAC system will depend on the above requirements, the environmental conditions (min/max temperature, humidity) at the plant location and the heat emissions from the equipment housed in the building (electrical cabinets and cables, mechanical equipment).

Heating, Ventilation & Air Conditioning (HVAC)

Climatic Data

Warm season

Design Temperature for Ventilation Systems	+ 26.2 °C
Design Temperature for Air Conditioning Systems	+ 30.8 °C
Absolute Maximum Temperature	+ 41.0 °C
Specific Enthalpy for Air Conditioning System Design	+ 66.8 kJ/kg
Relative Humidity	60 %

Internal Design Condition

Warm season

Rooms with permanent working personnel	+ 24 °C

For Technological Control Rooms the following optimal rates shall be maintained round a year:

Temperature	22 ± 2 °C
Relative Humidity	50 ± 10 %

ESTIMATED HEAT EMISSION FROM EQUIPMENT (W/m² OF FLOOR AREA)	
CONTROL ROOMS	350
OFFICES, LABORATORIES, CLINIC	-
ELECTRICAL SWITCH ROOMS	50
KITCHENS	250
DINING AREAS	50
MAINTENANCE AREAS	15

The HVAC of an industrial facility with a large number of buildings even in a non-extreme climate can be a significant electric power consumer. In such cases, HVAC power consumption impacts the sizing of the power generators.

In such cases, the buildings HVAC power consumption must be precisely evaluated at an early stage in the Project.

Heat emissions from equipment will not be available at this stage and will be estimated. A preliminary sizing of the HVAC equipment is done on this basis, resulting in an estimated power consumption. Such estimate is critical as a significant underestimate might lead to improper sizing of the power generators.

Chapter 8

Material & Corrosion

Materials & Corrosion discipline specifies the materials to suit the various service conditions. It also specifies how these materials will be protected against internal (from fluid) and external (atmospheric) corrosion.

Equipment and pipes material selection is done on the basis of required material strength (ability to withstand pressure), adequacy with fluid temperature and resistance to corrosion from the carried fluid.

The most common material encountered is Carbon steel, which is cheap and widely available. It comes in different grades. High strength grades are used for high pressure service, to reduce the wall thickness. For very low temperature, such as depressurization lines and cryogenic service, alloy steels, such as stainless steel, are required.

On Off-Shore platforms, where sea water is used for fire fighting, the fire water distribution ring is usually made of GRE (Glass Reinforced Epoxy). Highly alloyed steel, such as super duplex stainless steel, is used for strength at connections to fire water pumps.

Material & Corrosion

Material selection is done on the basis of the calculated corrosion rate.

Steel pipes handling well stream effluent in Oil and Gas production facilities, for instance, are subject to corrosion by acid water. Indeed, the effluent from the wells contains a mixture of oil, water and gas. Gas contains CO_2, which makes the water acid. Acid water corrodes steel.

The total corrosion rate, i.e., loss of thickness, over the design life of the facility is calculated, based on the CO_2 pressure, the fluid temperature, etc.

If such loss is only a few mm, then ordinary carbon steel "CS" is selected, with an increased thickness, called a corrosion allowance "CA", typically up to 6 mm only.

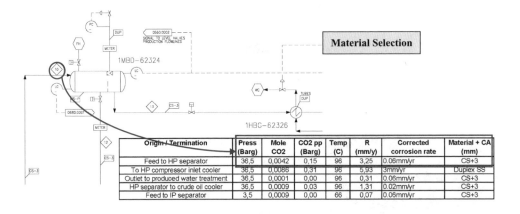

Material Selection

Origin / Termination	Press (Barg)	Mole CO2	CO2 pp (Barg)	Temp (C)	R (mm/y)	Corrected corrosion rate	Material + CA (mm)
Feed to HP separator	36,5	0,0042	0,15	96	3,25	0.06mm/yr	CS+3
To HP compressor inlet cooler	36,5	0,0086	0,31	96	5,93	3mm/yr	Duplex SS
Outlet to produced water treatment	36,5	0,0001	0,00	96	0,31	0.06mm/yr	CS+3
HP separator to crude oil cooler	36,5	0,0009	0,03	96	1,31	0.02mm/yr	CS+3
Feed to IP separator	3,5	0,0009	0,00	66	0,07	0.06mm/yr	CS+3

If the corrosion rate is high, a corrosion resistant alloy steel must be selected instead, such as stainless steel.

In some cases, it is possible to inhibit corrosion by injecting a chemical, called corrosion inhibitor, to decrease the corrosion rate. In such cases the pipes can remain in carbon steel but adequate corrosion monitoring, usually by means of weight loss coupons and corrosion probes, must be put in place to ensure inhibition is effective.

Carbon steel is not suitable where clean service is required and must be galvanized to ensure cleanliness.

Material selection is specified by the Material and Corrosion Engineer and shown on the **Material Selection Diagrams**.

Material & Corrosion

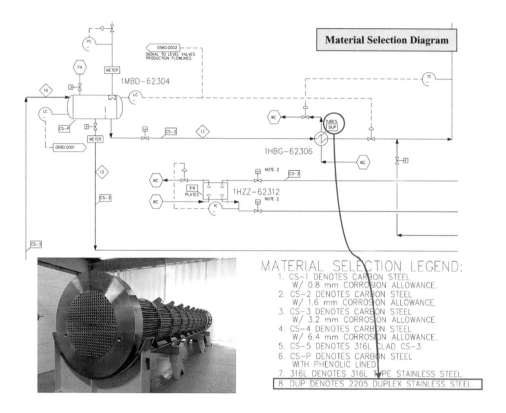

Materials have very different ability to withstand temperature and corrosion. Many of them, however, have the same visual apppearance. In order to avoid confusion, for instance use of the wrong type of alloy during piping fabrication, which could lead to catastrophic line rupture, adequate marking and inspection of materials is put in place.

Marking is specified in the piping material purchase specification. On top of that, Positive Material Identification (PMI) is carried out at Site for alloy steels. PMI determines the chemical composition of the steel, allowing to differentiate two identically looking alloys.

DAILY POSITIVE MATERIAL IDENTIFICATION REPORT FOR PIPING

ISOMETRIC No.: F65A775RBD1016 3R0JL	✓ FOR ACCEPT ✗ FOR REJECT	REPORT No : 7727539 PAGE No : 01

MATERIAL TYPE	WELD METAL TYPE	PMI EQUIPMENT :
A: 304L B: 304H C: 316L D: NiCrMo4 E: Other Alloy	A: 308L B: 308H C: 316L D: NiCrMo4 E: Other Alloy	NITON XLI/XLT

SPOOL NO.	W.No. BW / FW	BASE METAL 1	WELD METAL	BASE METAL 2	EXAMINED BY	DATE
	06	A	A	A	AK	30-03-09

Mo	Nb	W	Ni	Fe	Mn	Cr	V	Ti
0.24 ± 0.07	0.03 ± 0.03	0.01 ± 0.28	8.46 ± 1.39	68.87 ± 2.15	2.26 ± 1.01	18.59 ± 1.25	0.25 ± 0.46	0.26 ± 0.57
0.26 ± 0.06	0.00 ± 0.01	0.00 ± 0.22	8.98 ± 1.13	67.51 ± 1.73	2.05 ± 0.78	19.31 ± 1.02	0.00 ± 0.51	0.55 ± 0.52
0.23 ± 0.09	0.02 ± 0.03	0.00 ± 0.36	8.79 ± 1.81	71.85 ± 2.81	0.63 ± 1.10	17.69 ± 1.57	0.22 ± 0.55	0.00 ± 1.10
0.28 ± 0.06	0.01 ± 0.01	0.00 ± 0.24	8.29 ± 1.11	70.00 ± 1.75	2.64 ± 0.80	18.20 ± 0.99	0.01 ± 0.29	0.00 ± 0.65

The corrosion engineer also specifies the protection of structures and pipes against external (atmospheric) corrosion.

Protection of outdoor steel from corrosion is achieved by coating. The coating can be a metallic coating, such as Zinc (galvanizing) or Aluminium (very severe environment). For less severe requirements, steel is painted, after thorough surface preparation (sand blasting).

Painting is done following a painting system. The painting system defines the number of layers, the composition and the thickness of each layer. Different painting materials are used for pipes in low temperature and high temperature service.

The painting specification defines the surface preparation and paint system to be used for each application. Reference is made to an International code for the definition of the colors.

No.	Pipework Category	Painting System
1.	Pipes, factory bends, tees and other fittings with service temperature up to 80°C	*Epoxyvinyl System* Primer: inorganic zinc primer, DFT 75 µm min. Intermediates: two coats of epoxyvinyl paint, DFT 80+100 µm. Top coat for final color: epoxy paint, DFT 40 µm min. Total DFT 295 µm min.
2.	Pipes, factory bends, tees and other fittings with service temperature over 80°C	*Silicone System* Primer: inorganic zinc primer, DFT 75 µm min. Intermediates: two coats of silicone paint, DFT 25+25 µm. Top coat for final color: silicone paint, DFT 25 µm min. Total DFT 150 µm min.

Protection of submerged steel, e.g., internals of vessels, Off-Shore platform jacket, sealines, is done by means of sacrificial metallic attachments.

Such attachments, made of a less noble metal than steel, corrode first and, as they are electrically connected to the protected steel, prevent the corrosion of the latter. Sacrificial anodes are usually in zinc. They can be replaced once consumed.

Protection against corrosion of steel buried in the ground, e.g., underground piping services, is also achieved by coating. A mechanically stronger coating than painting is required for such application, usually in the form of a polymer applied at the factory on the straight pipes, fittings, etc. Field joints are coated at Site. The **coating specification** defines the requirements of the coating, such as surface preparation, number, material and thickness of layers.

Buried steel pipes are usually protected against corrosion by an additional system, called the **cathodic protection** system.

Cathodic protection consists of maintaining the steel pipe at a low negative potential. This is done by flowing an electric current between the pipe and an anode buried close to it. Anodes are surrounded by material of low resistance, such as coke, in order to ensure the flow of the electric current. Reference electrodes measuring the pipe potential are provided to control that the pipe is effectively protected.

Electrodes used to monitor pipe potential

Cathodic protection

Coke breeze

Laying of flexible anode

The **Insulation specification** covers the different types of insulation installed on equipment and piping: insulation for heat conservation, personnel protection as shown here and acoustic insulation. It specifies the insulation materials (such as mineral wool), thickness and provides detailed requirements for proper installation, ensuring in particular an adequate protection from the weather.

Chapter 9

Piping

Based on the Process Fluid list obtained from Process, and the material selection having been made, Piping discipline defines different groups (called classes) of piping material.

For instance:

- piping for all low pressure non corrosive fluids will be class A,
- piping for low pressure low corrosive service will be class B (extra thickness – called corrosion allowance – added compared to class A),
- piping material for low pressure highly corrosive service will be class C (material changed from carbon steel used in class A and B to stainless steel), etc.

Definition of the **Piping Material Classes** allows to standardize the piping material by using the same for several services. In this way, material will be interchangeable at Site. Any excess material for any lines of a given class can be used for any other line of the same class. Should there be a change at Site on one of these lines, it will be easier to find available material.

Process Fluids list

FLUID	SYMBOL	OPERATING CONDITIONS				MATERIAL
		T (°C)		MPa(abs)		
		MAXI	DESIGN	MAXI	DESIGN	
Drain	BD	30	50	atm	2	CS
Drain	BD	30	50	atm	9.95	CS
Drain	BD	50	70	atm	26.6	CS
Fuel Gas	FG	30	50	0.8	1.0	SS
Fuel Gas	FG	40	60	4.5	5.0	SS
Fuel Gas	FG	55	75	9.8	9.95	CS
Diesel fuel	FO	amb	50	0.2	0.4	CS
Fire Water	FW	amb	50	1.1	1.3	HDPE
Fire Water	FW	amb	50	1.1	1.3	CS
Lube Oil	LO	30	80	0.42	0.6	GALVAN
Methanol	ME	20	50	atm	0.4	SS
Methanol	ME	20	50	25.45	26.6	SS
Open drain	OY	amb	50	atm	0.4	CS
Hydrocarbon Gas	P	30	50	atm	2	CS
Hydrocarbon Gas	P	30	50	9.8	9.95	CS
Hydrocarbon Gas	P	-40/30	-46/50	atm	0.3	LTCS
Hydrocarbon Gas	P	-40/30	-46/50	9.8	9.95	LTCS
Hydrocarbon Gas	P	138	160	25.35	26.6	CS
Hydrocarbon Gas	P	50	70	25.35	26.6	CS
Hydrocarbon Gas	P	138	160	25.35	29.2	CS
Hydrocarbon Gas	P	-40/138	-46/160	25.35	29.2	LTCS
Hydrocarbon Gas	P	-40/50	-46/70	25.35	26.6	LTCS
Utility Air	UA	30	50	1.1	1.3	CS
Utility Water	UW	amb	50	0.3	0.5	GALVAN

Piping material classes

PIPING CLASS	RATING #	MATERIAL	COR. (mm)	TEMPERATURE		PRESSURE	
				MAX OPER °C	DESIGN °C	MAX OPER Barg	DESIGN Barg
11A	150	CS	0	30	50	atm	19
15A	600	CS	0	30	50	97	98,5
18A	2500	CS	0	138	160	253	265
21A	150	LTCS	0	-40/30	-46/50	atm	2
25A	600	LTCS	0	-40/30	-46/50	97	98,5
28A	2500	LTCS	0	-40/50	-46/70	253	265
28B	2500	LTCS	0	-40/138	-46/160	253	291
31A	150	SS	0	30	50	7	9
31U	150	SS	0	20	50	atm	3
35A	600	SS	0	40	60	44	49
38A	2500	SS	0	20	50	253	280
91A	150	CS-GALVA	0	30	80	3,2	5
91A	150	CS-GALVA	0	amb	50	2	4
10P	150	HDPE	0	amb	50	10	12

The pictures below show two types of "exotic" materials: Cu-Ni (Copper-Nickel alloy) and GRE (Glass Re-inforced Epoxy) used for high (fresh water) and very high (sea water) corrosive services.

Piping

Piping involves a large variety of components: straight runs, elbows, tees, flanges, reducers, valves, etc. Each of these components must be specified in order to be purchased. This is done in the **Piping Material Classes** specification.

Piping material classes

PIPING CLASS	RATING #	MATERIAL	COR. (mm)	TEMPERATURE MAX OPER °C	TEMPERATURE DESIGN °C	PRESSURE MAX OPER Barg	PRESSURE DESIGN Barg	ASME DESIGN CODE
11A	150	CS	0	30	50	atm	19	B31-8
15A	600	CS	0	30	50	97	98.5	B31-8
18A	2500	CS	0	138	160	253	265	B31-8
21A	150	LTCS	0	-40/30	-46/50	atm	2	B31-8
25A	600	LTCS	0	-40/30	-46/50	97	98.5	B31-8
28A	2500	LTCS	0	-40/50	-46/70	253	265	B31-8
28B	2500	LTCS	0	-40/138	-46/160	253	291	B31-8
31A	150	SS	0	30	50	7	9	B31-8
31U	150	SS	0	20	50	atm	3	B31-3
35A	600	SS	0	40	60	44	49	B31-8
38A	2500	SS	0	20	50	253	280	B31-8
91A	150	CS-GALVA	0	30	80	3,2	5	B31-3
91A	150	CS-GALVA	0	amb	50	2	4	B31-3
10P	150	HDPE	0	amb	50	10	12	B31-3

Piping material class specification

SERVICE : DRAIN (BD) HYDROCARBON GAS (P)
GENERAL MATERIAL : CARBON STEEL — API 5L Gr. B, X52, X65
Corrosion Allowance = 0
RATING : 2500# RTJ
PIPING CLASS : 18A
CODE: ASME B31-8
Page: 1/3

Limits:
T °C	-29	38	121	160
P Barg	265	278	278	265

	DIA from	DIA to	Sched./ WT(mm) Rating	End	Material standard	Dimensions standard	DESIGNATION	NOTES
PIPE	1/2"	3/4"	160	BE	API 5L Gr.B-MDS-CS01	ASME B36.10	SEAMLESS PIPE	
	1"	1½"	XXS	BE	API 5L Gr.B-MDS-CS01	ASME B36.10	SEAMLESS PIPE	
	2"	2"	160	BE	API 5L Gr.B-MDS-CS01	ASME B36.10	SEAMLESS PIPE	
	3"	3"	80	BE	API 5L Gr.X52-MDS-CS04	ASME B36.10	SEAMLESS PIPE	4
	4"	14"	120	BE	API 5L Gr.X52-MDS-CS04	ASME B36.10	SEAMLESS PIPE	4
	16"	24"	(*)	BE	API 5L Gr.X65-MDS-CS06	ASME B36.10	S.A.W. WELDED PIPE : (*) 16" thk = 25.4, 18" thk = 28.58 20" thk = 31.75, 24" thk =38.1	4
FORGED STEEL FITTING — B.W	1/2"	2"		BW	ASTM A105-MDS CS01	MSS SP-97	WELDOLET (BW AS PER ASME B16-25)	1
	3"	14"		BW	A694- F52-MDS CS03	MSS SP-97	WELDOLET (BW AS PER ASME B16-25)	1
	16"	24"		BW	A694- F65-MDS CS05	MSS SP-97	WELDOLET (BW AS PER ASME B16-25)	1
BUTT WELDING	1/2"	3/4"	160	BW	A234-WPB-MDS SC01	ASME B16.9	45°, 90°ELBOW, TEE, RED. TEE, CAP, REDUCER	
	1"	1½"	XXS	BW	A234-WPB-MDS SC01	ASME B16.9	45°, 90°ELBOW, TEE, RED. TEE, CAP, REDUCER	
	2"	2"	160	BW	A234-WPB-MDS SC01	ASME B16.9	45°, 90°ELBOW, TEE, RED. TEE, CAP, REDUCER	
	3"	3"	80	BW	MSS SP-75 WPHY 52-MDS CS03	ASME B16.9	45°, 90°ELBOW, TEE, RED. TEE, CAP, REDUCER	
	4"	14"	120	BW	MSS SP-75 WPHY 52-MDS CS03	ASME B16.9	45°, 90°ELBOW, TEE, RED. TEE, CAP, REDUCER	
	16"	24"	pipe thk	BW	MSS SP-75 WPHY 65-MDS CS05	ASME B16.9	45°, 90°ELBOW, TEE, RED. TEE, CAP, REDUCER	1
FLANGES	1/2"	2"	2500# RTJ	BW	ASTM A105-MDS CS01	ASME B16.5	WELDING NECK FLANGE	1
	3"	12"	2500# RTJ	BW	A694-F52-MDS CS03	ASME B16.5	WELDING NECK FLANGE	1
	14"	14"		BW	A694-F52-MDS CS03		HUB CONNECTOR (BW AS PER ASME B16-25)	1-2-3
	16"	24"		BW	A694-F65-MDS CS05		HUB CONNECTOR (BW AS PER ASME B16-25)	1-2-3
	2"	2"	2500# RTJ	BW	ASTM A105-MDS CS01	ASME B16.36	2 ORIFICE WN FLANGE + 1/2"PLUG+JACK SCREW	1
	3"	12"	2500# RTJ	BW	A694-F52-MDS CS03	ASME B16.36	2 ORIFICE WN FLANGE + 1/2"PLUG+JACK SCREW	1
	1/2"	12"	2500# RTJ	-	ASTM A105-MDS CS01	ASME B16.5	BLIND FLANGE	
	14"	24"			A694-F52-MDS CS03		BLIND HUB CONNECTOR	2-3
GASKET	1/2"	12"	2500#		SOFT IRON (90 HB max)	ASME B16.5,B16.20	OCTAGONAL RING- JOINT GASKET	
	14"	24"			AISI 4140		SEAL RING FOR HUB CONNECTOR (FOR CLAMP-TYPE DEVICE)	2-3
BOLTING					A193Gr.B7+ Zn Bichr.	ASME B16.5	STUD BOLT & 2 HEAVY HEX. NUTS	
					A194Gr.2H+ Zn Bichr.	ASME B1.1	DIA ≤1"COARSE Series, DIA >1" 8 THREADS series	
						ASME B1.1	SPECIAL BOLTING FOR CLAMP-TYPE DEVICE	3

Such specification defines:

- the metallurgy, specified by reference to an international material standard, e.g., API-5L,

SERVICE : DRAIN (BD) HYDROCARBON GAS (P)	GENERAL MATERIAL : CARBON STEEL API 5L Gr. B, X52, X65	RATING : 2500# RTJ	PIPING CLASS : 18A
	Corrosion Allowance = 0		Page : 1/3
			CODE: ASME B31-8

Limits				
T °C	-29	38	121	160
P Barg	265	278	278	265

		DIA		Sched./ WT(mm) Rating	End	Material standard	Dimensions standard	DESIGNATION	NOTES
		from	to						
	PIPE	1/2"	3/4"	160	BE	API 5L Gr.B-MDS-CS01	ASME B36.10	SEAMLESS PIPE	
		1"	1"½	XXS	BE	API 5L Gr.B-MDS-CS01	ASME B36.10	SEAMLESS PIPE	
		2"	2"	160	BE	API 5L Gr.B-MDS-CS01	ASME B36.10	SEAMLESS PIPE	4
		3"	3"	80	BE	API 5L Gr.X52-MDS-CS04	ASME B36.10	SEAMLESS PIPE	4
		4"	14"	120	BE	API 5L Gr.X52-MDS-CS04	ASME B36.10	SEAMLESS PIPE	4
		16"	24"	(*)	BE	API 5L Gr.X65-MDS-CS06	ASME B36.10	S.A.W. WELDED PIPE : (*) 16" thk = 25.4 , 18" thk = 28.58 20" thk = 31.75 , 24" thk =38	4
FORGED STEEL FITTING	B.W.	1/2"	2"		BW	ASTM A105-MDS CS01	MSS SP-97	WELDOLET (BW AS PER ASME B16.25)	1
		3"	14"		BW	A694- F52-MDS CS0:			
		16"	24"		BW	A694- F65-MDS CS0:			
	BUTT WELDING	1/2"	3/4"	160	BW	A234-WPB-MDS SCC			
		1"	1"½	XXS	BW	A234-WPB-MDS SCC			
		2"	2"	160	BW	A234-WPB-MDS SCC			
		3"	3"	80	BW	MSS SP-75 WPHY 52-MDS			
		4"	14"	120	BW	MSS SP-75 WPHY 52-MDS			
		16"	24"	pipe thk	BW	MSS SP-75 WPHY 65-MDS			
	FLANGES	1/2"	2"	2500# RTJ	BW	ASTM A105-MDS CS			
		3"	12"	2500# RTJ	BW	A694-F52-MDS CS03			
		14"	14"		BW	A694-F52-MDS CS03			
		16"	24"		BW	A694-F65-MDS CS05			
		2"	2"	2500# RTJ		ASTM A105-MDS CS			
		3"	12"	2500# RTJ		A694-F52-MDS CS03			
		1/2"	12"	2500# RTJ		ASTM A105-MDS CS			
		14"	24"	-		A694-F52-MDS CS03			
	GASKET	1/2"	12"	2500#		SOFT IRON (90 HB n			
		14"	24"			AISI 4140			
	BOLTING					A193Gr.B7+ Zn Bichr A194Gr.2H+ Zn Bichr			

CHEMICAL REQUIREMENTS FOR HEAT ANALYSES (Section 6)

Type of pipe		Grade	Carbon maxi % (1)	Manganese maxi % (1)	Phosphorus maxi %	Sulfur maxi %
seamless		A	0.22	0.90	0.030	0.030
Non-expanded or cold expanded		B(6)	0.27	1.15	0.030	0.030
		X42	0.29	1.25	0.030	0.030
Non-expanded		X46(4), X52(4),	0.31	1.35	0.030	0.030
Cold expanded		X42(4), X46(4), X52(4),	0.29(2)	1.25	0.030	0.030
Non-expanded or cold expanded		X56(3,4), X60(3,4),	0.26	1.35	0.030	0.030
		X65, X70, X80			(by agreement)	

TENSILE REQUIREMENTS (Section 6)

Grade	Yield strength minimum		Ultimate tensile strength minimum		Ultimate tensile strength maximum		Elongation minimum percent in 2 in. (50.8 mm)
	ksi	MPa	ksi	MPa	ksi	MPa	
A	30.0	207	48.0	331			
B	35.0	241	60.0	413			
X42	42.0	289	60.0	413			
X46	46.0	317	63.0	434			See note (1)
X52	52.0	358	66.0	455			
X56	56.0	386	71.0	489			
X60	60.0	413	75.0	517			

- the geometry/dimensions, by reference to international dimensional standard, e.g., ASME B16.9 for the elbows (defining the length, etc.),

Piping

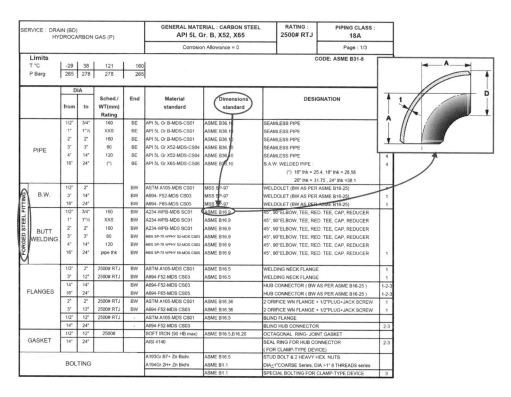

- the wall thickness (by steps, called schedules, for standardization reasons), for each diameter, which is calculated as per the applicable design code, pressure, temperature, material properties and allowance for corrosion.

Piping wall thickness calculation

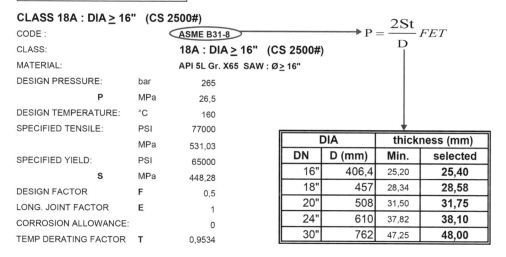

Piping

Every piping item, besides straight lengths, must be counted in order to be purchased: elbows, tees, flanges, etc.

Using the P&IDs (for item count) and the Piping Layout drawings (for lengths), a preliminary list of the required piping material, called first **Piping Material Take-Off (MTO)**, is prepared. Typically, the first MTO will focus on long lead time items (large diameter, unusual materials) rather than standard off-the shelf ones (ordinary carbon steel, usual diameter). The first purchase order of piping material is placed for typically 70% of the quantities of the first MTO. This allows to start procuring piping material to ensure first supply to Site at an early date, while avoiding wastage: as only 70% is ordered, no excess material will have been ordered even if quantities decrease by up to 30% during detailed design.

Subsequent revisions of the piping Material Take-Off will be automatically generated by the 3D design software (see corresponding section). The designer will route each line in the 3D model, by selecting items (straight runs, elbows) from a catalog, for the corresponding piping class. The software will then produce the consolidated list of items, for all lines.

The specification of valves is a major task of the Piping engineer. It includes very detailed material requirements for valve internals, i.e., type of alloy required for trim, material of gaskets, etc. Indeed, these moving parts are subject to severe operating conditions (erosion, compression) and require specific material selection.

SERVICE : HYDROCARBON GAS (P)				GENERAL MATERIAL : ASTM A333 Gr. 6	RATING : 2500# RTJ	PIPING CLASS : 28A
				Corrosion Allowance = 0		
	DIA					
	from	to	Rating	End	DESIGNATION	Standard	Valve data sheet / Comments
BALL VALVE	1/2"	1"1/2	2500#	BW	Full bore with BW Nippples, Trunnion ball, 3-piece body Body:LTC.Steel Trim: 17/4 PH impact tested at -46°C Seats/Seals: PEEK / Viton or equal	API 6D ASME B16-34	VB 01-1-28A
	2"	20"	2500#	BW	Full bore, welded body with BW pup pieces Trunnion ball Body:LTC.Steel Trim: 17/4 PH impact tested at -46°C Seats / Seals PEEK or PTFCE / PTFE	API 6D ASME B16-34	VBF 01-2-28A
	2"	20"	2500#	BW	Reduced bore, welded body with BW pup pieces Trunnion ball Materials: Same as above	API 6D ASME B16-34	VB 01-2-28A

The Piping material specialist will review the valve vendors drawings and check that the material offered for the valve body, trim, gaskets, etc. are compliant or equivalent to the ones specified in the specification and valves data sheets.

Piping routing studies starts from the Process Flow Diagrams, which show which interconnections are to be made between equipment, and the Plot Plan, which shows the location of these equipment.

Line Diagrams, also called "line shoot diagrams", are produced from Process and Utility flow diagrams, showing all pipes going from a particular plant area to another, regardless of the pipe services. These are geographical drawings, compared to Process and Utility flow Diagrams which are functional diagrams.

They allow to identify the required location and width of the main pipe ways/racks.

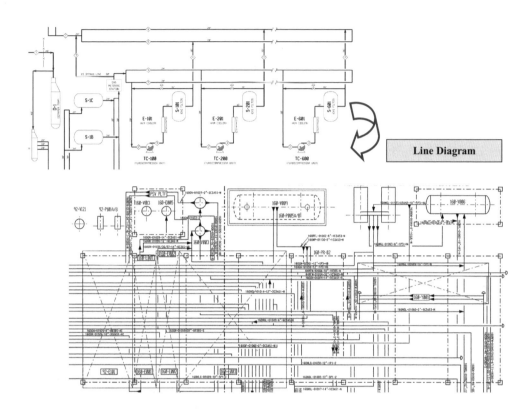

Piping installation studies are then done to precisely define the routing of pipes, e.g., from pipe-rack to equipment, etc.

The installation specialist needs to have a general view encompassing a large number of requirements:

- Process requirements: a fluid in gravity flow, for instance, will require a sloped line routing, very low pressure drop allowance will require a very short routing, etc.
- Equipment details: elevation of nozzles for pressure vessels, dictated by process and shown on equipment guide drawings.

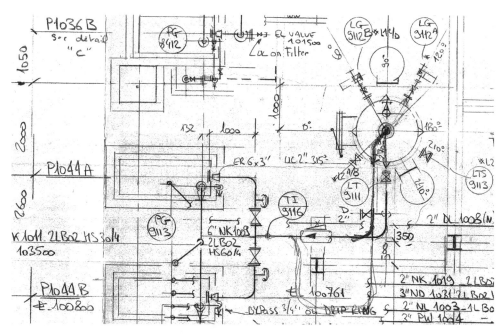

- Piping flexibility: provision of direction changes or expansion loop in the line to allow its expansion while subject to temperature change in operation[1],
- Grouping with other lines on common support/pipe-rack,
- Structural design constraints, such as width of piperack, span between bays, etc.,
- Space for dismantling and handling of equipment for maintenance: provision of clearance on top of equipment for

---

1. The straight routing of a line between two vessels would cause excessive forces on the vessels when the line temperature increases due to thermal expansion. Instead, a routing with a number of direction changes, or even a purpose-made expansion loop, must be adopted to allow the line to expand.

Detailed stress calculations are performed at a later stage in the design, to confirm that the routing of critical lines (high temperature, etc.) provides enough flexibility. The key is to be able to guess a correct routing, before these calculations are done. Indeed, they go into many details (location of supports, etc.) so that it would not be feasible to test a number of possible routings. The calculations will then simply validate the guessed routing.

hoisting equipment, clearance for heat exchanger bundle pull-out, lay down area for dismantled parts, vehicle/forklift access where required, etc.
- Operator access: hand wheels of valves, instruments must be accessible to operators and therefore be placed at a suitable elevation. These ergonomics consideration are essential for the plant safe operation and are called Human Factor Engineering.

The above requirements translate into typical piping arrangements around control valves, pumps, which ensure sufficient working area for operational and maintenance tasks. For a pump, it would for instance include easy access to drain valves, space to remove the inlet strainer, etc.

Piping

Piping studies set the widths of pipe ways, the requirements for piping and equipment support structures, such as the elevation of concrete support structures for equipment, the width and number of levels of pipe-racks, etc.

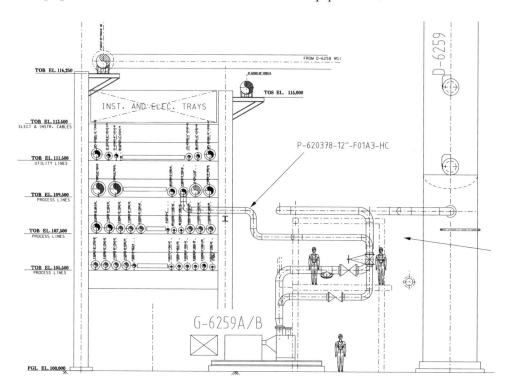

The 2D routing of the main lines is drawn on the **Piping Layout drawings**.

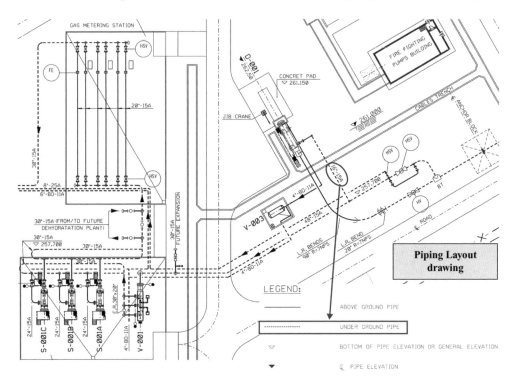

As explained above, the Piping Layout drawings allow to determine pipe lengths and to produce the first Piping Material take-off.

Besides layout drawings, Piping issues construction drawings, which are of two types: the **General Arrangement Drawings**, which are used for piping erection, and the **Isometric Drawings**, which are used for piping pre-fabrication.

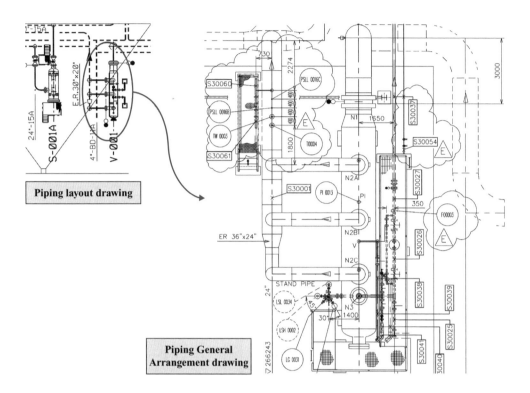

The **Piping General Arrangement drawings** are very detailed. They contain all information necessary for erection of piping: all dimensions, elevations, position of in-line items, etc. They were originally produced to allow the production of Isometric drawings, when the latter was a manual task. As a the 3D model software is now used to produce Isometrics, Piping General Arrangement drawings are no longer systematically produced.

They were also used to give a view of the complete environment within an area, showing all equipment, pipes, valves, structures, etc. They tend nowdays to be replaced by snapshots taken from the 3D model.

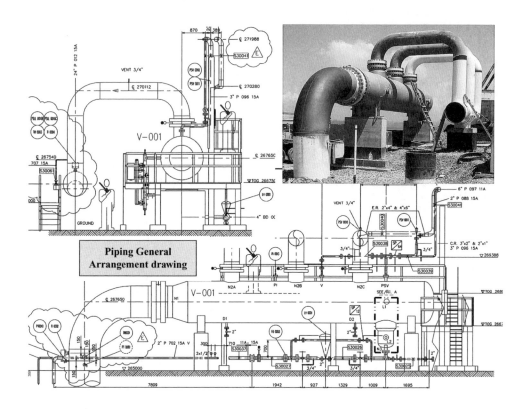

Piping General Arrangement drawing

Piping Isometric Drawings show a 3D view of an individual line, with all the dimensions defining its geometry and the list of all its components (straight length, elbows, etc.).

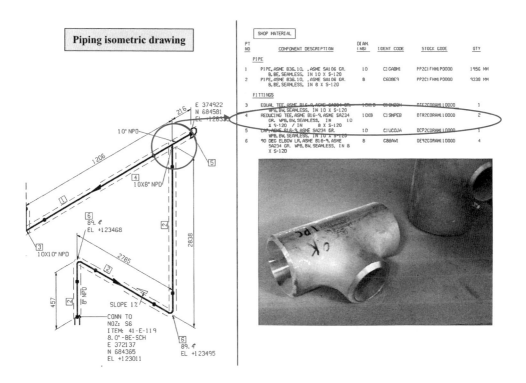

Piping isometric drawing

As piping materials are very numerous and resemble each other, e.g., it is not easy to identify one steel alloy from another, identifiers are stamped by the piping material supplier on each item and the identifiers are shown on the Isometric drawings.

The Isometric drawing produced by Engineering is not directly used for construction. Indeed, as the line will be pre-fabricated in parts, called spools, drawings must be issued showing how the line is divided into spools. Shop isometric drawings are issued to this end by the Construction contractor. They are also used to identify welds, each of which will be associated with inspection and test records.

Piping

Design ISO **Shop ISO**

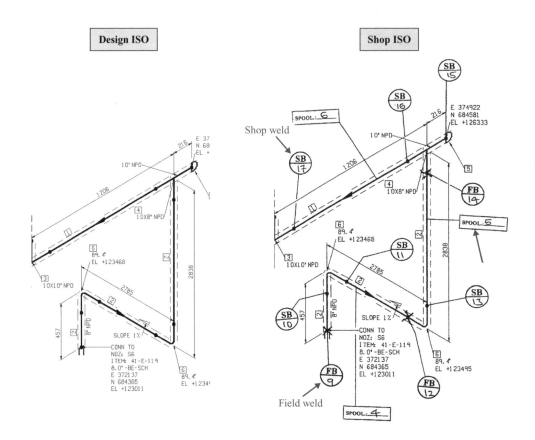

Shop weld

Field weld

Piping Pre-fabrication: Gas cutting, welding

The fabrication workshop (shop) Isometrics also specify all fabrication requirements, such as the welding procedure to be used, the required inspection (surface of weld or in-depth inspection by means of radiography for instance), special operation such as heat treatment of welds, etc.

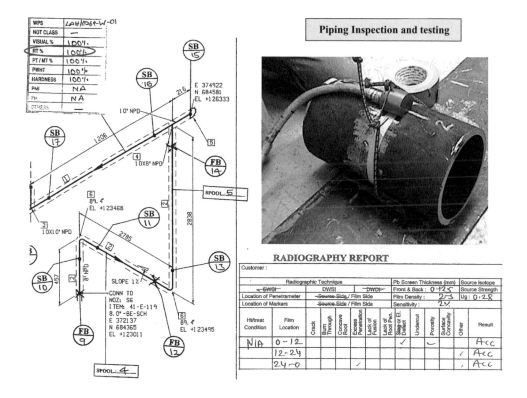

The **Line List** produced by Process is completed by Piping, with construction requirements, such as:

- specific non-standard fabrication requirements, such as heat treatment of welds, for high thickness pipes or pipes in corrosive service,
- specific testing requirements of piping material (for alloy steel, to ensure that the material used is the right alloy),
- inspection requirements (also called NDE: Non Destructive Examination) for welds: type, e.g., radiographic examination for gas services or surface defect control only for less critical services, extent of the inspection, e.g., 100% of welds to be tested for gas service, 10% only for non-critical service, etc.

Piping

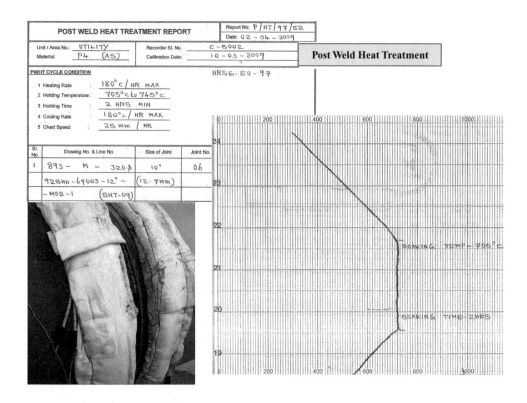

- pressure test requirements including type (hydraulic, pneumatic test or service test only), depending on criticality of the line service, and pressure,
- type of coating,
- type of insulation (heat conservation, personnel protection, cold insulation) and thickness, etc.

Fluid Code	Unit Code	Seq No.	Line Size	Class	Line Connection From	Line Connection To	Paint Code	Insulation Code	Insulation Thk.	PWHT	Hardness	NDE Butt Welds	NDE Fillet Welds	Pressure Test Medium	Pressure Test Press (barg)	PMI
GN	71	61106	22	3C3AS1	LNG STORAGE	UNIT 93	1C	N	NO	YES	YES	A,B	A	H	51,80	0%
GN	71	61106	20	3C3AS1	LNG STORAGE	UNIT 93	1C	N	NO	YES	YES	A,B	A	H	51,80	0%
GN	71	61106	12	3C3AS1	LNG STORAGE	UNIT 93	1C	N	NO	YES	YES	A,B	A	H	51,80	0%
LNG	71	60001	32	3R0JLL	668-P001 A/B/C	LNG RUNDOWN HEADER	7S	6	180	NO	NO	A,D,F	A,F	P	33,00	100%
LNG	71	60001	22	3R0JLL	668-P001 A/B/C	LNG RUNDOWN HEADER	7S	6	170	NO	NO	A,D,F	A,F	P	33,00	100%
DOW	72	63000	0,75	1P1	72-P061A	DOW	1C	N	NO	NO	NO	A,B	A	H	3,00	0%
DOW	72	63001	0,75	1P1	72-P061B	DOW	1C	N	NO	NO	NO	A,B	A	H	3,00	0%
DOW	72	63002	0,75	1P1	72-P062A	DOW	1C	N	NO	NO	NO	A,B	A	H	3,00	0%
DOW	72	63003	0,75	1P1	72-P062B	DOW	1C	N	NO	NO	NO	A,B	A	H	3,00	0%

These inspection and testing requirements are shown on the Piping Isometric drawing in order to make it a document encompassing all information required for construction.

SERVICE CLASS	INSULATION CODE	PAINTING CODE	TEST FLUID	RT (%)	PT or MT (%)	PMI (%)	PWHT	INTERNAL CLEANING	TEMP. DGN. (°C)	TEMP. OPE. (°C)	PRESS. DGN. (Bar.)	PRESS. OPE. (Bar.)	P&ID NO.
1S1	1	50	H	5	0 0		NO		195	155	7.0	4.1	

It is a very tedious exercise to prepare the numerous piping isometrics (several thousands on a typical size job). Therefore this process is automated using the computer aided 3D design software, in which all the plant pipework is modelled.

The routing of lines done by the Piping designer takes into account good engineering practise and provides for flexibility in the line.

Nevertheless, this routing must be checked by the **Piping Stress Analysis & Supports** group. This is done by calculation of the stress in the line during operation (due to thermal expansion, pressure, etc.) and checking that the stress is within the maximum allowable for the concerned material and at the connection to equipment nozzles. The line installation temperature, i.e., the outside temperature prevailing at Site during the installation of the line, is one input of the calculation. The line thermal expansion is indeed due to the difference between this temperature and the line operating temperature.

A computer model allowing finite element analysis is used for this purpose.

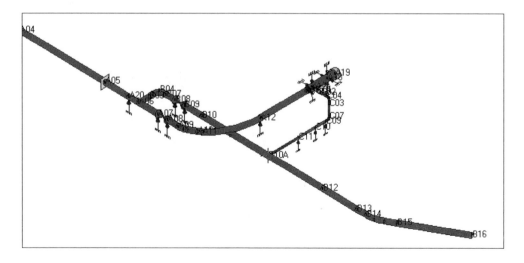

The results, which confirm that the routing is acceptable, are recorded in **Stress Calculation Note**.

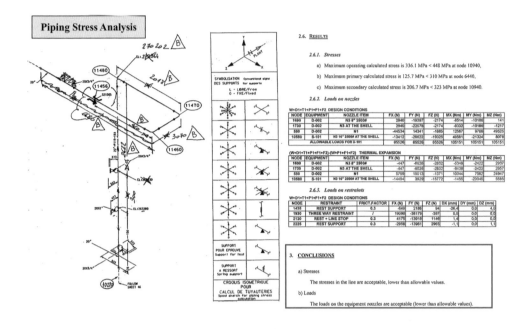

Once the routing is confirmed to be acceptable, the piping Isometric drawings are issued for construction.

Piping Stress Analysis is systematic for high temperature/ high pressure service and large diameter lines. When such a line comes into service, if it cannot expand, because it is constrained, it will be subject to internal stress, which could exceed the strength of the steel or make the line come off its position and even fall

from its supports. Allowance for displacement of the line must be provided in the design, by means of expansion loop and guiding supports. Typically, an expansion loop, such as the one depicted here, is provided to absorb the expansion of the line between two anchor (fixed) points.

Special attention is paid at connections to rotating equipment, where the flexibility of the inlet & outlet lines must ensure that minimum loads are transferred to equipment nozzles.

Indeed, excessive forces on the connected equipment could result in its displacement. Typically, the driven equipment, a pump, for instance, will be displaced, resulting in its misalignment with the driver (motor). A special pipe routing, such as the one depicted below, is implemented to reduce the loads on the machinery flanges to prevent this.

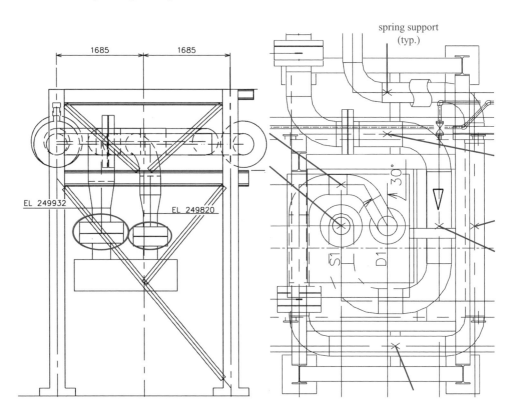

Spring supports are provided to minimize the connecting lines loads on equipment nozzles.

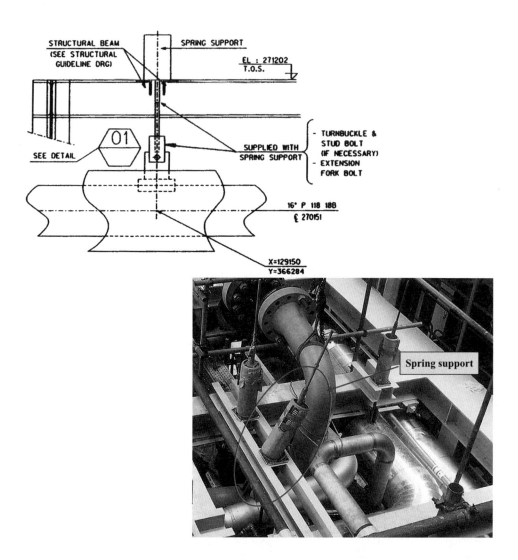

Positions of supports must also allow, during installation of the pipe work, effortless fitting of the pipes connecting to nozzles of rotating machinery. Very stringent tolerances are imposed by vendors in terms of parallelism between pipe flanges and rotating machinery nozzles. Forced connection of piping flanges to machine nozzles using hydraulic tools can generate forces and moments on nozzles and induce shaft misalignment. This can create serious problems to the machine.

Pipe supports are otherwise standardized to the maximum extent, in order to allow their mass pre-fabrication.

Piping

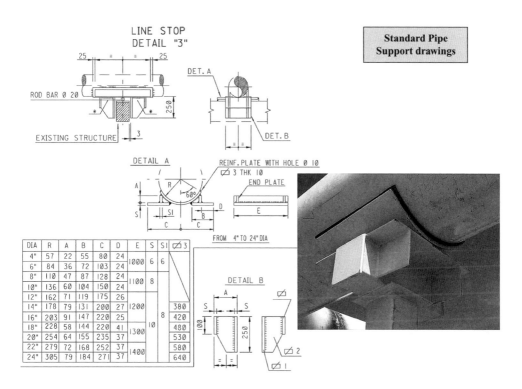

A few pipe supports, called **Special Pipe Supports**, are non standard, for which individual drawings are issued by the Piping Support group.

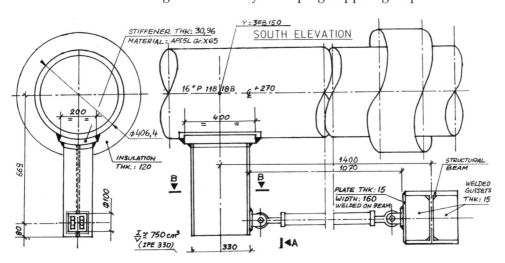

Chapter 10

Plant model

◆

Plants, specially Off-Shore platforms, are usually congested due to the limited space available. Several disciplines install their equipment in the same limited space: equipment, pipes, supports, structural steel, cables, etc. This must be coordinated in order to avoid interferences, e.g., pipe and structural steel members installed at the same place, etc.

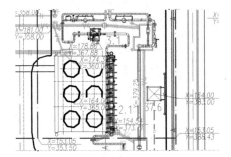

This coordination used to be done in 2D, by superimposing the various discipline location drawings that were at the time and for that reason done on transparencies, e.g., piping, foundations, underground piping, cable routing plans, all having the same coordinate system, etc.

Superimposing drawings then became a functionality of 2D design softwares such as AutoCAD, which allow the various disciplines to work in independent, yet superimposed, layers identified by different colors on the screen, e.g., cable sleeves in green, pipes in black… At any time in its design, the piping engineer can display the civil layout in order to check for civil interference with its own design.

Plant model

Computer Aided Design systems are now in 3 dimensions, allowing to build a **3D model** of the plant. Models of plants used to be made using glue and plastic parts. This is now replaced by virtual (digital) 3D models, which are stored on a server and can be accessed by many users at the same time and from different locations.

All significant materials are modelled to scale. The model reflects exactly what the plant will be. All buildings, roads, escape ways, structures, equipment, pipes, valves, cable trays, junction boxes, etc. are modelled in details by each engineering discipline.

The use of a 3D model is particularly useful for Off-Shore platforms, where space is limited and its use shall be optimized.

Using such a system allows to identify and clear interferences between disciplines in congested areas. Besides manual visual review of possible interferences in the model, the system can perform automatic **clash checks**, in order to pinpoint the interferences left unnoticed.

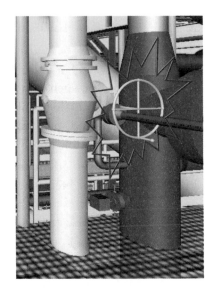

Model reviews of the virtual plant are carried out at various stages in the design (typically 30%, 60% and 90% progress).

Such reviews allow to check operator accesses, overhead clearances, non-obstruction of escape ways, free space for equipment dismantling during maintenance, etc.

Ideally all disciplines, e.g., structural steel, piping, etc. not only inputs its design objects (steel structure beams and columns, pipework, valves, etc.) in the model but also performs its design in the model itself, rather than on specific discipline models, and issues directly its construction drawings from the model. This ensures that the latest information is in the model, e.g., if a steel beam depth has been recently changed from 1,000mm to 1,200mm, the pipe router will see it immediatly and be able to locate its pipe at the right elevation so that the latter will not clash with the steel beam.

Items modelled include one-off items, such as a pressure vessel, a package, a motorized/control valve, an in-line strainer, and standard items, such as a steel section, a piping elbow, etc. which are part of a catalogue. Using a catalogue allows to define each standard item, complete with detailed dimensions and specification, only once. This information will then appear on all occurrences of this item.

Modelling of virtual objects is also done, such as volumes reserved for escape ways, travel of dismantled equipment/parts during maintenance, etc.

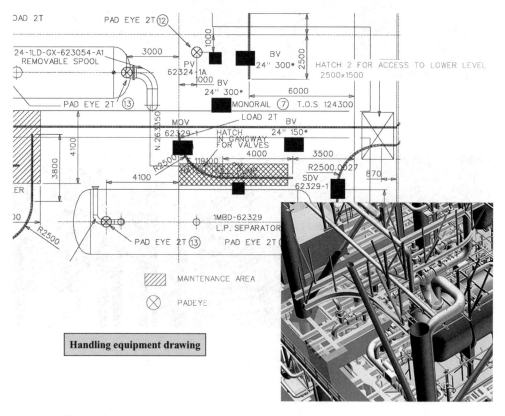

Handling equipment drawing

Modelling of equipment is first done with estimates of equipment dimensions. Indeed, actual dimensions of equipment, which are sized by vendors, are not known initially.

Once vendor information becomes available, the equipment model is up-dated based on vendor drawings: exact dimensions, shapes, nozzle orientation, etc.

When modelling vendor packages, it is very important to model all items of the package, e.g., not only the main equipment, but also structural steel, package internal piping, etc. and to up-date these models with revisions of vendor drawings.

A register of items modelled, complete with indication of reference and revision of the vendor drawing, is maintained in each discipline to this end.

Modelling is not only done for large equipment, but also for smaller ones, such as motorized valves, particularly in Off-Shore environment where space is limited. Dimensions of motorized valves, including their actuators which can be very big, are non standard. Those dimensions will not be known before sizing has been done by the valve vendor.

As engineering progresses, each discipline adds its own objects and networks to the model, e.g., secondary structure, access platforms, stairways, ladders, piping supports, cable trays, instruments, junction boxes, lighting, etc.

In such a way, each designer finds the best location/routing for its items, according to required access, available free space, etc. It also allows to minimize the cost by finding the shortest routes for pipes, cables, etc.

Once equipment have been modelled and main pipe ways have been defined, lines are modelled in the 3D model. As discussed above, piping is purchased in individual components: straight pipes, elbows, tees, reducers, flanges. Each of these components must be modelled. This could be a very lengthy exercise. Fortunately, it is made easy by standardization of piping material and creation of a catalogue of items in the 3D model.

Lines are modelled using the items from the catalogue for the corresponding piping class. This allows a very fast "just pick and place" modelling, provided one has populated the catalogue with all items before hand.

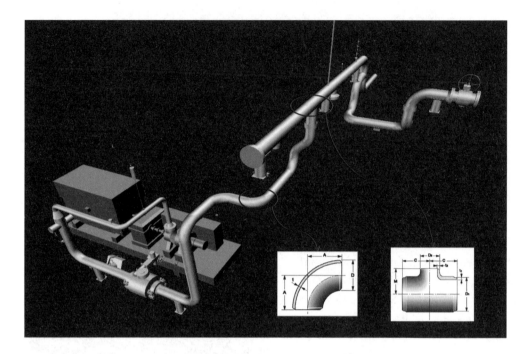

When extracting the piping Isometric drawing from the 3D model, dimensions and specifications of all items will come automatically from the catalogue.

REV	Item	SIGMA CODE CLIENT CODE	large diam.	small diam. /length	QUANTITIES TO BE SUPPLIED		BALANCE TO BE SUPP. A - B
					New A	Old B	
		SEAMLESS PIPE					
		TE04170	AP15LGRB	-ANSIB36-10 -BW -SCH40 -			
005	34		P	2	1248	1212	36
		TE04900	AP15LGRB	-ANSIB36-10 -PLAIN END -SCH80 -			
004	37		P	3/4	6	6	0
004	36		P	1	6	6	0
004	42		P	11/2	6	6	0
		TE14795	AP15LGRB	-ANSIB36-10 -PLAIN END -SCH80 -MDS CS01 -			
005	44		P	1/2	138	0	138
005	25		P	3/4	12	6	6
004	24		P	1	6	6	0
004	43		P	11/2	6	6	0

Once piping modelling is advanced enough, a revision of the **Piping Material Take-Off (MTO)** can be extracted from the 3D model. Balance is made between this latest bill of required material and the material already ordered on the basis of the first MTO. Shortage material is purchased through an amendment to the purchase order. The MTO will in fact be extracted several times from the model, in order to purchase piping material as early as possible. Routing of large diameters lines and lines in exotic materials will be prioritized as corresponding material has long lead time. As soon as the corresponding lines are routed in the model, their MTO will be extracted and the material ordered. The piping MTO will therefore undergo several revisions.

The modelling activities of the various engineering disciplines take place, and associated documents are issued, as depicted on the flowchart that follows.

Plant model

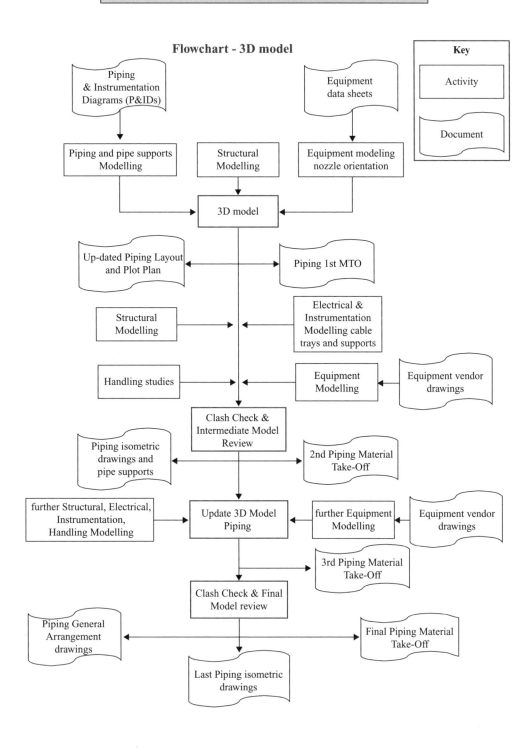

Similarly to Piping, other disciplines extract bill of quantities from the 3D model, such as steel structures, cable trays, etc.

On Off-Shore projects, the model is also used to determine the global module/platform weight and centre of gravity, from the individual components weight and centre of gravity.

Chapter 11

Instrumentation and Control

◆

Instrumentation design starts from the P&IDs, on which all required instruments, controls and automations have been shown by the Process discipline covering:

- process monitoring,
- process control,
- process safety (alarm, shutdown).

Process not only defines and shows on the P&IDs the required function: process value to be measured (Pressure, temperature, flow) but also the required function (indication, recording, control) and whether the information shall be available locally (like pressure is, for instance, on the gauge shown here), in a local instruments panel located in the field next to the equipment, or remotely in the control room.

Instrumentation and Control

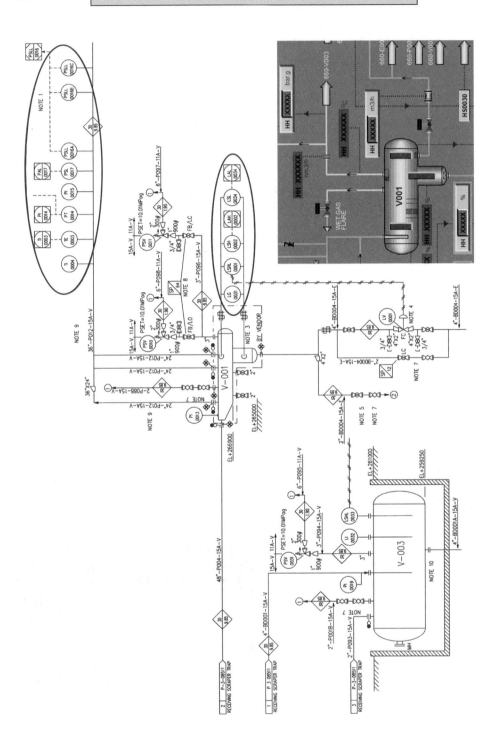

Instrumentation and Control

Instrumentation discipline implements the requirements specified by Process:

- specifying and ordering the systems, and developing their detailed functional specifications,
- specifying and ordering field instruments, and all accessories necessary for their installation, i.e., accessories for instrument process and electrical connections,
- producing all the drawings required for equipment and instruments installation and wiring.

Monitoring and control of the process is performed by the Process Control System (PCS). Process controls consist of both logic (ON/OFF) and analog (continuous) controls.

Control requirements, e.g., temperature in such vessel shall be controlled by varying flow of cooling medium using such control valve, are defined by Process, shown on the P&IDs and described in the Operating & Control philosophy.

Functional requirements are specified by means of **Functional diagrams**. One such diagram is issued for each type of control: these are typical diagrams. The one shown here is for a split-range control, by the process control system, coupled with an emergency shutdown function, by the emergency shutdown system.

Instrumentation and Control

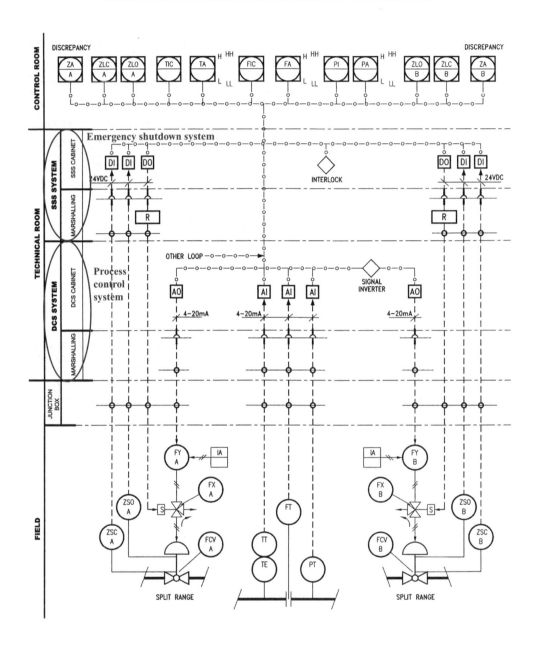

Specific and complex controls are described to the control system supplier in **Control Narratives**.

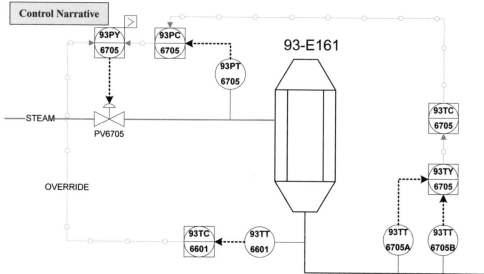

Temperature is measured by two transmitters 93TT6705A/B. Operator selects the transmitter by 93HS6705 and a ramp is performed during switchover. When one transmitter is in bad value, controller used the value from the healthy one.

Controller 93PC6705 acts on valve 93PV6705. If temperature measured by 93TT6601 (93-E161 outlet) is very low (output of 93TC6601 will increase), 93PC6705 will be overridden by 93TC6601. This in order to prevent low temperature at 93-E161 outlet (93TT6705A/B are close to GF distribution utility area). Set point of controller 93TC6601 will be lower than set point of controller 93TC6705.

The specification of the system will entail gathering all the requirements in the **System specification**, and producing a number of other documents describing the system capacity, geographical spread and functionnalities.

Instrumentation and Control

The **System Architecture drawing** shows the various pieces of hardware of the system, their location, and the interfaces with other systems, including the electrical control system and the equipment control systems supplied by vendors.

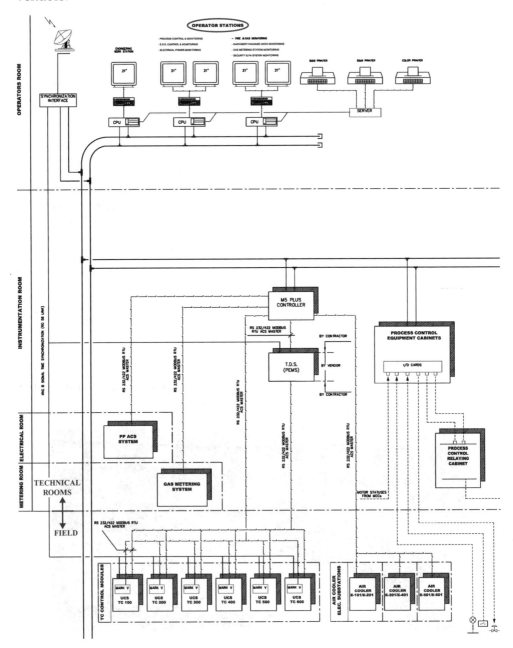

Instrumentation and Control

Marshalling cabinets and programmable controllers are located in instrumentation buildings/rooms spread throughout the plant. Indeed, they must be located close to the field instruments, to reduce cable lengths. Operator interface units (consoles) are centrally located in the control room.

The **I/O count** determines the required capacity of the system.

1) DISCRETE INPUT /OUTPUT LIST

POS.	DESCRIPTION	DI	DO	AI	AO	RTD
1	FIELD INSTRUM.	300	20	150	20	40
2	VALVES	280	60	-	-	-

In addition a +10 % spare Input /output shall be considered and additionally +20% space for future requirements.

I/O COUNT

2) SERIAL INPUT /OUTPUT LIST

POS.	DESCRIPTION	DI	DO	AI	AO
1	TC-100	200	-	50	-
2	TC-200	200	-	50	-
3	TC-300	200	-	50	-
4	TC-400	200	-	50	-
5	TC-500	200	-	50	-
6	TC-600	200	-	50	-
7	FIRE & GAS	1200	-	-	-
8	GAS METERING	60	20	60	10
9	POWER SUPPLY	100	-	30	-

The system engineer specifies the **Mimic displays** to the control system vendor, i.e., the content of the views that will be displayed on the operator consoles.

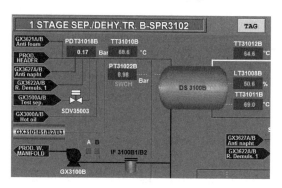

Such displays are the Plant Operator's interface with the control system. Their adequacy is critical. They are reviewed with the Client's operations staff.

Process emergency shutdown is performed by the Emergency ShutDown (ESD) system. The ESD system is a separate system from the Process Control system. This ensures redundancy and independence. The ESD system has its functions internally duplicated or triplicated to ensure high reliability.

The ESD system initiates process equipment shutdown and closure of isolation valves in an emergency. The shutdown logic is implemented in the ESD system as defined by Process on the ESD Cause & Effects diagrams.

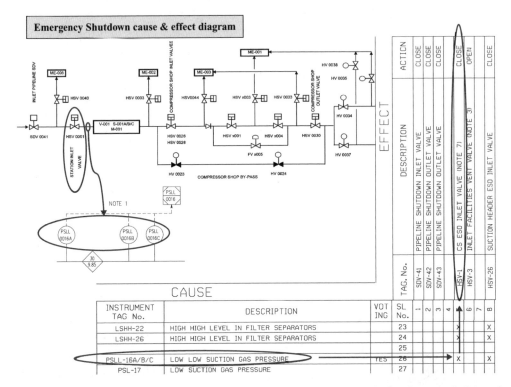

A **SIL (Safety Integrity Level) review** is carried out to check the reliability of critical safety automations. Such a critical automation is, for instance, the closure of an isolation valve in case of a major leak.

Each automation is allocated a severity level based on the consequence that would result from its failure.

The severity is ranked, for instance level 1 would be loss of production, level 2 damage to equipment, level 3 release of flammable substance to atmosphere, level 4 would be loss of containment, etc.

The frequency of occurrence times the severity, the risk level, determines the required level of reliability, for instance failure on demand between 10^{-3} to 10^{-4}.

The reliability of the automation foreseen to be installed is then estimated using information from the instrumentation hardware (transmitter, I/O card, system). Should it prove below the required reliability, additional/redundant components are added to increase its reliability.

Instrumentation and Control

In the example shown above, three low pressure sensors had to be provided to secure the closure of the isolation valve upon detection of a major leak/line rupture.

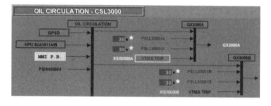

The ESD system vendor programs the automation logic in the system. The system displays the status of activation of emergency levels and implementation of actions to the operator.

All plant instruments are logged in a master register: the **instrument index**. This data base is progressively filled with all information: service conditions (P,T), instrument type, signal output, material of construction, range, set point, etc.

The instrument data base centralizes all information pertaining to each instrument. Many documents (wiring diagrams, loop diagrams, etc.) and list of materials (hook-up) are produced directly from this unique data base, ensuring their consistency.

Tag Number	Instrument Type	Location	Service	Equipment/Line	PID N°	I/O Type	Signal	System
AE -0701-1	Analyse measure	MAH	Gas metering station		P-3-08540	AI	4-20 mA	
AT -0701-1	Analyser transmitter	SBMR	Gas metering station		P-3-08540	AI	SL	GMS
AI -0701-1	Analysis indicator	SBMR	Gas metering station		P-3-08540	-	Soft	GMS
AXA -0703-6	Apparatus failure alarm	SBMR	Gas metering station		P-3-08540	DO	24 Vdc	GMS
ASHH -0703-2	Very High dew point switch	SBMR	Gas metering station		P-3-08540	DO	24 Vdc	ESD
AT -1061	Moisture analyser	Field	Pilot gas system TC-100	S-105	P-3-08555	-	-	-
BE -1201-1	Flame detector	Field	Power turbine TC-100		NUO/10.07/00171	AI	UV	UCS (TC-100)
BL -1201-1	Flame indicator	CMTC-100	Power turbine TC-101		NUO/10.07/00171	-	Soft	UCS (TC-100)
BXA -1201-1	Flame detector fault alarm	CMTC-100	Power turbine TC-102		NUO/10.07/00171	-	Soft	UCS (TC-100)
FT -0013	Flow transmitter	Field	Fuel gas for turbocompressors	8"-FG001-15A-V	P-3-08541	AI	4-20 mA	PCS
FO -1003	Restriction orifice	Field	TC-100 Emergency vent	4"-P107-28A-V	P-3-08514	-	-	-
FE -1005	Orifice plate	Field	TC-100 Suction	20"-P101-18A-B	P-3-08516	-	-	-
FT -1005	Flow transmitter	Field	TC-100 Suction	20"-P101-18A-B	P-3-08516	AI	4-20 mA	PCS

Instrument index

Tag Number	Instrument Type	I/O Type	Signal	System
AE -0701-1	Analyse measure	AI	4-20 mA	-
AT -0701-1	Analyser transmitter	AI	SL	GMS
AI -0701-1	Analysis indicator	-	Soft	GMS
AXA -0703-6	Apparatus failure alarm	DO	24 Vdc	GMS
ASHH -0703-2	Very High dew point switch	DO	24 Vdc	ESD
AT -1061	Moisture analyser	AI	UV	UCS (TC-100)
BE -1201-1	Flame detector	-	Soft	UCS (TC-100)
BL -1201-1	Flame indicator	-	Soft	UCS (TC-100)
BXA -1201-1	Flame detector fault alarm	-	Soft	UCS (TC-100)
FT -0013	Flow transmitter	AI	4-20 mA	PCS
FO -1003	Restriction orifice	-	-	-
FE -1005	Orifice plate	-	-	-
FT -1005	Flow transmitter	AI	4-20 mA	PCS
FY -1005-1	Computing device	-	Soft	PCS
GD -0108	Gas detector (IRGD)	AI	4-20 mA	CF-001
GAH -0108	Gas alarm (10%LEL)	-	-	CF-001
HZLH -2002-1	Status valve opened	Hold	Soft	UCS (TC-200)
JQ -0737	Energy flowrate totalizer	-	Soft	GMS
JI -0738	Energy flowrate indicator	-	Soft	GMS
LSHL -0001	Pneum level switch	-	3-15 psi	-
LV -0001	Control valve	-	3-15 psi	-
SH -0002	High level switch	DI	24 Vdc	PCS
LAH -0002	High level alarm	-	Soft	PCS
MXS -0730-3	Remote position switch	DI	24 Vdc	CF-004
MXL -0730-3	Local/remote status	-	24 Vdc	CF-004
MXS -0730-4	Start order	DO	220Vdc	CF-004
MXS -0731-1	Electrical fault switch	DI	24 Vdc	CF-004
ND -0034	Fire detector(UV/IR)	DI	24 Vdc	CF-001
NAH -0034	Fire detection alarm	-	Soft	CF-001
PSLL -0049-C	Very low pressure switch	DI	24 Vdc	ESD
PALL -0049	Very Low pressure alarm	-	Soft	ESD
PSH -0050	High pressure switch	DI	24 Vdc	PCS
PAH -0050	High pressure alarm	-	Soft	PCS
ST -4602	Speed transmitter	AI	Analog	BNR
SI -4602	Speed indicator	AO	Analog	TDS
TT -0707	Temperature transmitter	AI	Hard	GMS
TI -0707	Temperature indicator	-	Soft	GMS
TAL -0707	Low temperature alarm	-	Soft	GMS
TAH -0707	High temperature alarm	-	Soft	GMS
VT -1201-2	Vibration transmitter	AI	Analog	BNR
VI -1201	Vibration indicator	Hold	-	-
WI -1102-1	Torque indicator	-	Soft	UCS (TC-100)
WI -1102-2	Torque indicator	-	Soft	UCS (TC-100)
ZAHH -2602	Very high displacement alarm	-	Soft	UCS (TC-200)
ZE -2604-1	Displacement detector	Hold	-	-

A **data sheet** is produced for each instrument, specifying the range, material of construction, etc. in order to purchase it, as well as for reference for its maintenance at site.

Instrumentation and Control

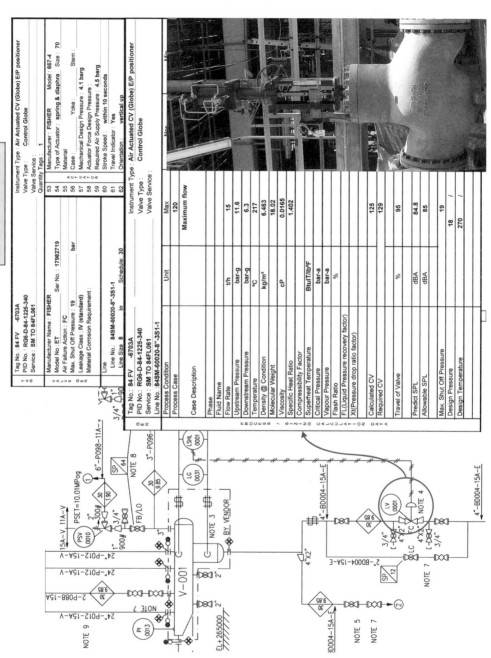

Control valve data sheet

Tag No.: 84 FV -6703A	Instrument Type : Air Actuated CV (Globe) E/P positioner	
PID No.: RG6-D-84-1225-340	Valve Type : Control Globe	
Service : SM TO 84FL061	Valve Service :	
	Quantity Tags : 1	
Manufacturer Name : FISHER	Manufacturer : FISHER Model: 667-4	53
Model No.: ET	Type of Actuator: spring & diaphra Size : 70	54
Air Failure Action : FC	Material : Stem :	55
Max. Shut Off Pressure : 19 bar	Case : Yoke :	56
Leakage Class : IV (standard)	Mechanical Design Pressure : 4.1 barg	57
Material Corrosion Requirement :	Actuator Force Design Pressure :	58
Line	Required Air Supply Pressure : 4.5 barg	59
Line No.: 84SM-60020-8"-3S1-1	Stroke Speed : within 10 seconds	60
Line Size 8 in Schedule 30	Travel Indicator : Yes	61
	Orientation : vertical up	62

Tag No.: 84 FV -6703A		Instrument Type : Air Actuated CV (Globe) E/P positioner	
PID No.: RG6-D-84-1225-340		Valve Type : Control Globe	
Service : SM TO 84FL061		Valve Service :	
Line No.: 84SM-60020-8"-3S1-1			

	Unit	Max	
Process Condition		120	
Process Case		Maximum flow	
Case Description			
Phase			
Fluid Name			
Flow Rate	t/h	15	
Upstream Pressure	bar-g	11.6	
Downstream Pressure	bar-g	6.3	
Temperature	°C	217	
Density @ Condition	kg/m³	6.463	
Molecular Weight		18.02	
Viscosity	cP	0.0165	
Specific Heat Ratio		1.402	
Compressibility Factor			
Superheat Temperature	BtuT/lb°F		
Critical Pressure	bar-a		
Vapour Pressure	bar-a		
Flash Ratio	%		
FL (Liquid Pressure recovery factor)			
Xt (Pressure drop ratio factor)			
Calculated CV		125	
Required CV		129	
Travel of Valve	%	95	
Predict SPL	dBA	84.8	
Allowable SPL	dBA	85	
Max. Shut Off Pressure		19	/
Design Pressure		18	/
Design Temperature		270	

The specification of level instruments requires the Instrument engineer to perform a level study based on the liquid levels specified by process. Level instruments deduct the liquid level by measuring the weight of the liquid column. Such weight is obtained from the difference of pressure at the bottom and on top of the column. A level instrument therefore includes a pair of pressure sensors: one located above the level to be measured and one located below.

Defining the elevation of the two pressure connections so that they adequately sense the weight of the liquid column is the purpose of the level study. It results in the **level sketches**, which serve to specify the level instruments and to define the elevations of the nozzles on the vessel. These elevations are specified to the vessel vendor. Level sketches are also issued to the Piping/Installation discipline, which addresses the requirements for access to instruments, e.g., definition of adequate access platforms/ladders, etc.

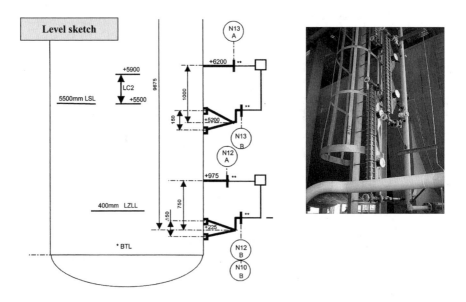

Instrumentation and Control

Instrumentation equipment and materiels, from the field sensor to the control room, are shown on the synoptic below.

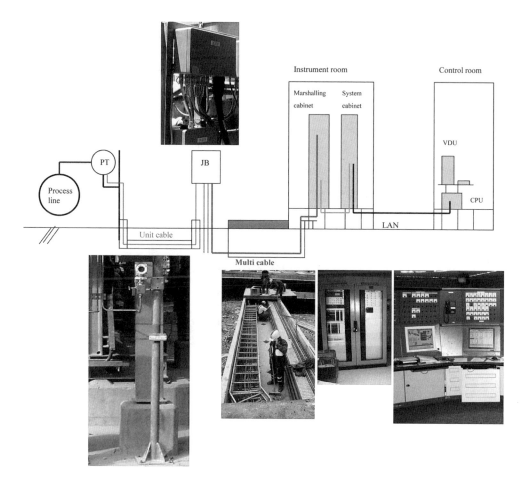

Instrumentation produces all drawings required for installation of these equipment and materials at Site, which include:

- The Junction Box Location drawings, which show the location of the junction boxes[1].

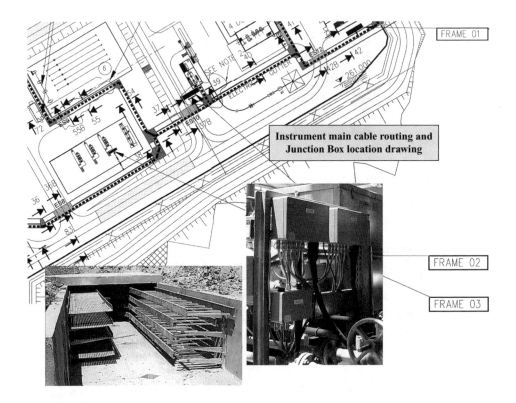

Instrument main cable routing and Junction Box location drawing

---

1. In order to reduce the number of cables connecting field instruments to cabinets in technical rooms, multi-core cables are used. They connect several instruments (typically 7/12/19), located nearby in the field, to the cabinet located in the instrumentation room. Instruments are connected to multi cables by means of junction boxes. Grouping of instruments in multi-core cables is done according to the nature of their signal (analog, digital, voltage level) and service/system (process monitoring, emergency shutdown).

- **Instrument location drawings,** which are derived from Piping General Arrangement drawings and show the location, position and elevation of fields instruments.

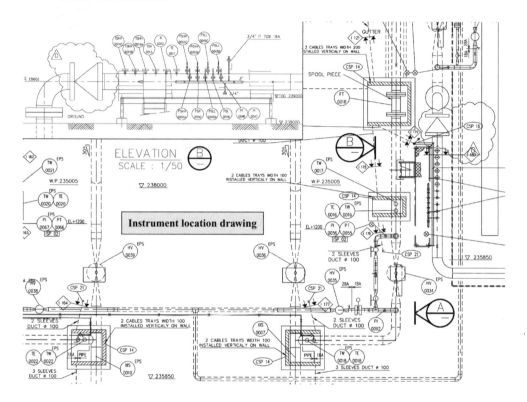

- **Cable routing drawings** showing in which cable trench/duct the cables shall be installed and **Cross section drawings** showing on which cable tray each cable shall be installed, in compliance with segregation rules, e.g., control cables and power supply cables on different trays,

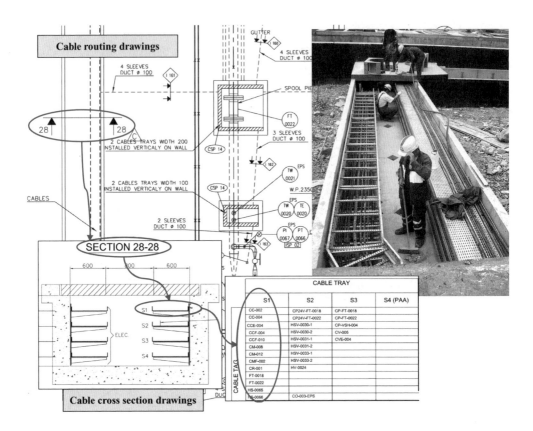

- **Instrument cable schedule**, showing the list of cables to install, cable type, length, origin, destination and route,

CABLES TAG	CABLES TYPE	SUPPLY BY (1)	FROM	LOCATION FRAME or OTHER	TO	LOCATION FRAME or OTHER	LENGTH m	ROUTING CROSS SECTIONS
CC-004	A-T-1-19-P-2-0	CONTRACTOR	JC-004	PIG D-002	CA-052	INSTRUM. ROOM	370	I121-I119-I161-28-11B-11-9-44-4-2-1-96
CC-005	A-T-1-12-P-2-0	CONTRACTOR	JC-005	FILTER-SEPARATOR	CA-052	INSTRUM. ROOM	440	I127-36B-36-36A-71-35-35B-65-34-3-1-96
CC-006	A-T-1-12-P-2-0	CONTRACTOR	JC-006	FILTER-SEPARATOR	CA-052	INSTRUM. ROOM	440	I127-36B-36-36A-71-35-35B-65-34-3-1-96
CC-007	A-T-1-7-P-2-0	CONTRACTOR	JC-007	STATION INLET VALVES	CA-052	INSTRUM. ROOM	370	I193-I190-36A-71-35-35B-65-34-3-1-96
CC-008	A-T-1-7-P-2-0	CONTRACTOR	JC-008	DIESEL GENERATOR	CA-052	INSTRUM. ROOM	120	I350-30-70B-70
CC-009	A-T-1-7-P-2-0	CONTRACTOR	JC-009	FIRE WATER	CF-004	FIRE BUILDING	100	I185-I183-I182-I181
CC-101-1	A-T-1-19-P-2-0	CONTRACTOR	JC-101	AERO E-101	UA-101	S/S ELECTRICAL 27-1	160	I207-I205-I204-I202-I200-24-62-27
CC-101-2	A-T-1-7-P-2-0	CONTRACTOR	JC-101	AERO E-101	UA-101	S/S ELECTRICAL 27-2	160	I207-I205-I204-I202-I200-24-62-27

Instrumentation and Control

The **Cable Material Take-Off** sums up the length of all cables, by type, showing the overall quantities to purchase.

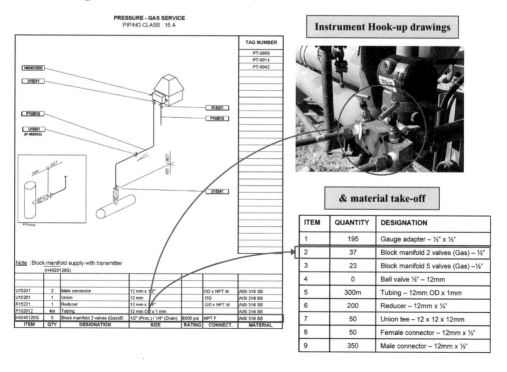

Cable Material Take-Off

- Instrument **Hook-up drawings**, which show mounting and connection of instrument to process lines and corresponding list of required material (tubing, manifold, connectors, etc.),

Instrument Hook-up drawings

& material take-off

ITEM	QUANTITY	DESIGNATION
1	195	Gauge adapter – ½" x ½"
2	37	Block manifold 2 valves (Gas) – ½"
3	23	Block manifold 5 valves (Gas) – ½"
4	0	Ball valve ½" – 12mm
5	300m	Tubing – 12mm OD x 1mm
6	200	Reducer – 12mm x ¼"
7	50	Union tee – 12 x 12 x 12mm
8	50	Female connector – 12mm x ½"
9	350	Male connector – 12mm x ½"

Instrumentation and Control

The **Bulk Material Take-Off** indicates the quantity of junction boxes, cable trays, small installation accessories (cable glands, cable markers, etc.), hook-up material, etc. to be purchased.

DESIGNATION	MATERIAL	RAW QUANTITIES	CONTINGENCIES	QUANTITIES TO BE PURCHASED
CABLE TRAYS (Return flange). (d: 50mm / w: 100mm) - Note 1	HOT-DIP GALVANIZED	956m	10%	1100m
CABLE TRAYS (Return flange). (d: 50mm / w: 200mm) - Note 1	HOT-DIP GALVANIZED	419m	10%	500m
CABLE TRAYS (Return flange). (d: 50mm / w: 400mm) - Note 3	HOT-DIP GALVANIZED	690m	10%	800m
CABLE TRAYS (Return flange). (d: 75mm / w: 600mm) - Note 1	HOT-DIP GALVANIZED	6000m	10%	8800m
COVERS FOR CABLE TRAYS. (w: 100mm)	HOT-DIP GALVANIZED	700m	10%	770m
COVERS FOR CABLE TRAYS. (w: 200mm)	HOT-DIP GALVANIZED	419m	10%	470m
COVERS FOR CABLE TRAYS. (w: 400mm) - Without junction boxes frames	HOT-DIP GALVANIZED	660m	10%	750m
COVERS FOR CABLE TRAYS. (w: 600mm)	HOT-DIP GALVANIZED	2100m	10%	2400m
HORIZONTAL TEES. (Return flange). (d: 75mm / w: 3x600mm)	HOT-DIP GALVANIZED	65	10%	75
90° HORIZONTAL BEND. (Return flange). (d: 75mm / w: 600mm)	HOT-DIP GALVANIZED	67	10%	75
COVERS FOR TEES. (w: 3x600mm)	HOT-DIP GALVANIZED	17	10%	20
COVER FOR 90° HORIZONTAL BEND. (w: 600mm)	HOT-DIP GALVANIZED	17	10%	20

For junction boxes, the MTO specifies the number of terminals, the number and diameter of cables (for cable entries in the JB), the size of the cores (for sizing of terminals, etc.). An **arrangement drawing**, such as the one shown here, may be attached to the junction boxes requisition to provide more detailed or specific requirements.

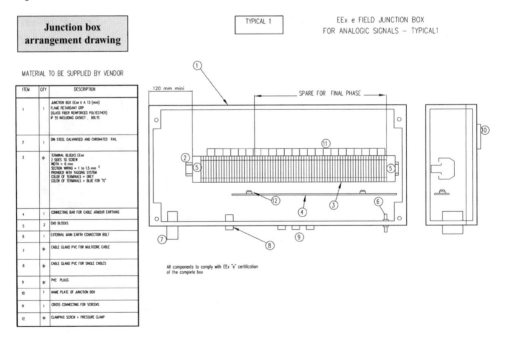

- **Standard installation drawings**, such as instrument, junction box and cable tray support drawings, earthing drawings, etc., which show typical arrangements,

Instrumentation and Control

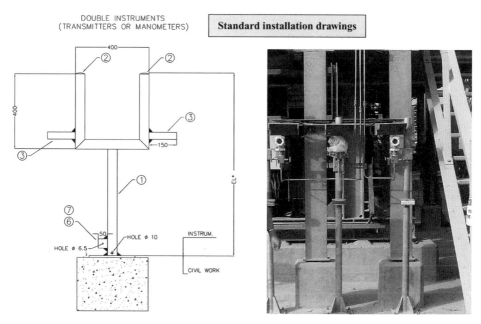

Standard installation drawings

- Equipment Layout drawings, showing arrangement of cabinets inside instrumentation technical/control rooms,

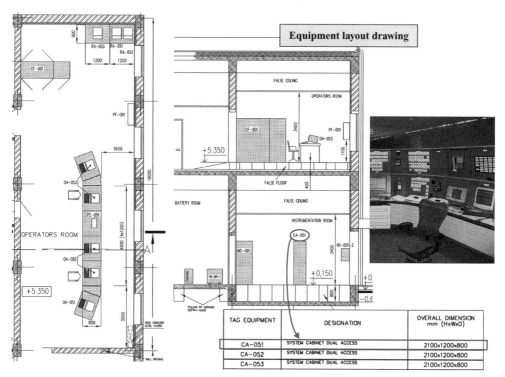

Equipment layout drawing

TAG EQUIPMENT	DESIGNATION	OVERALL DIMENSION mm (HxWxD)
CA-051	SYSTEM CABINET DUAL ACCESS	2100x1200x800
CA-052	SYSTEM CABINET DUAL ACCESS	2100x1200x800
CA-053	SYSTEM CABINET DUAL ACCESS	2100x1200x800

- **Wiring diagrams** show cable connections at terminals of junction boxes and marshalling cabinets,

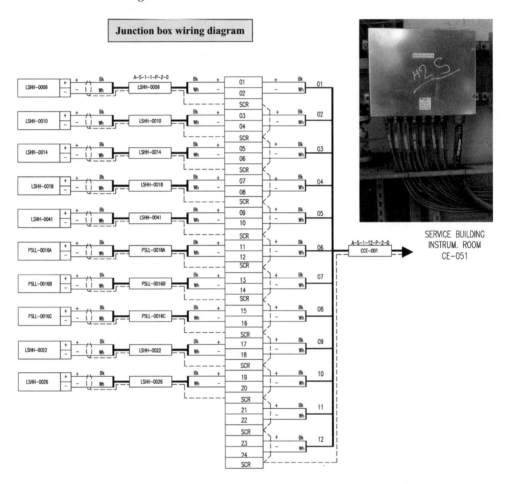

- **Loop Diagrams**, also called troubleshooting diagrams, show the complete wiring of each instrument. They are used during the testing of the instrument (from the field to the display on screen) during commissioning and for maintenance,

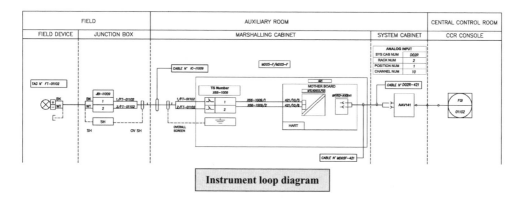

Instrument loop diagram

The lists of tagged items, such as the instrument index, cable schedule, etc. are used for the inspections and tests, prior the hand-over to the client, as part of Mechanical Completion activities. The type of inspection required depends on the type of item: calibration for instruments, insulation test for cables, etc. Each inspection is recorded against the item inspected.

A computer software "the mechanical completion system" is used to record the requirements and status of the inspection and testing of the thousands of individual tagged items.

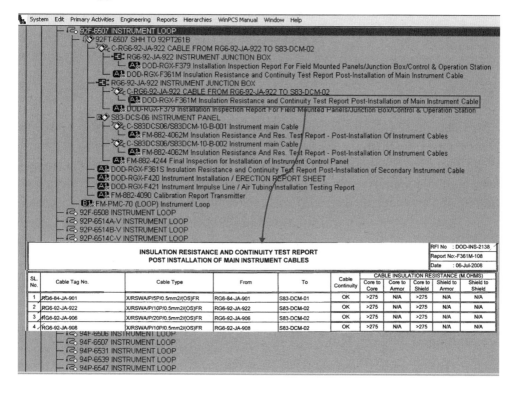

As stated in the Equipment section, sub-functional units of the plant are often purchased as "packages", already assembled and wired.

In such case, the instruments of the packages come with the package. The control of the unit may be performed in a dedicated local system, supplied by the package vendor, or integrated within the plant central process control system.

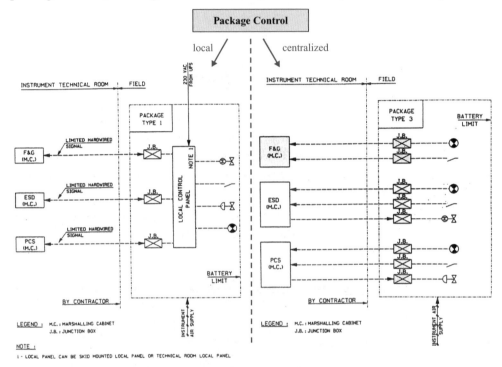

If the control system is supplied with the package, it will consist of a local control panel to which the package instruments will be wired. The local control panel performs the control actions for the package. The main process control system of the plant is simply interfaced to this panel to allow centralized monitoring and control, e.g., start-up/shut down, etc.

In order not to have too many different types of control systems it is usually preferable to have the package controls performed by the plant central control system.

In this case, the package instruments are wired to junction boxes at the skid edge. The junction boxes are connected to the plant Process Control System marshalling cabinets as all other field instruments and the controls/automations are configured in the system.

It is in such case critical that the control functions and automations are properly described by the package vendor: the lube oil pump must be started before the turbine is ignited!

To this end the package vendor provides the **Control Philosophy, Cause & Effect charts, Logic analysis and Schematics** to describe in details all sequence (start-up, shut down, etc.).

Similarly to the Process control system, Instrumentation discipline implements a **Fire and Gas detection and alarm system**. This is generally a similar system to the ESD system. The functional requirements are given by Safety (see corresponding section). Instrument discipline specifies and procures the materials (detectors, sounders, etc.), the system, and produces all drawings for Site installation.

The system is purchased based on the required capacity (I/O count). It is also specified to interface with the stand-alone Fire & Gas detection and Fire fighting systems of the main equipment packages, and with the plant ESD system. The system vendor programs the logic shown on the F&G matrix (see Safety section) in the system.

The same deliverables are produced for the Fire and Gas system as for the Process Control System: instrument list, location drawings, cable schedule, bill of materials, wiring and troubleshooting diagrams, etc.

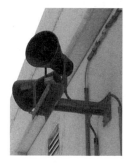

Other systems fall in the scope of the Instrumentation engineer, such as the Public address system (for paging personnel or sounding general alarm using loudspeakers, etc.), the plant internal telephone system (PABX), the computer network (LAN), the access control system, CCTV, etc.

An Off-Shore facility requires telecommunication with land, supply boats, tankers, etc. This will involve a variety of systems, which will be designed by the Telecommunication engineer, such as radio frequency (UHF, VHF), microwave, satellite, entertainment system (TV) in living quarters, etc.

Chapter 12

Electrical

Electrical engineering is in charge of the design of the plant electrical power generation and distribution.

Similarly to Instrumentation, the activities of the Electrical discipline can be categorized as follows: architecture (of the electrical power generation and distribution system), specification of all equipment and materials, and production of installation drawings.

Electrical engineering activities start with the identification of all consumers. This is done from the Equipment list and shall also include all electrical consumers "hidden" inside packages, such as a machinery lube oil heater, HVAC of buildings – sometimes a major load -, outdoor lighting, building lighting and small power, etc. All electrical consumers are registered in the **Electrical Load List**.

Equipment actual power consumption is not available initially, as the equipment make and model is not known yet. Electrical discipline estimates the power consumption first. The estimate is then replaced by the actual power consumption once the equipment (make and model) has been selected.

Once the consumers are identified, the total electrical power requirement of the plant can be evaluated. This is not the sum of the power requirements of all consumers. Indeed, they do not all operate simultaneously. A more refined approach is required to work out the realistic overall power demand.

LV Motor Control Center TG-MCC-002 (V=400/230V)							
Equipment No.	Description	Vital	Essential	Normal	Restarting	Duty Type	ABSORBED LOAD (A) kVA - kW

Equipment No.	Description	Vital	Essential	Normal	Restarting	Duty Type	ABSORBED LOAD (A)
TG-002	Turbo Gen.						
88CR	Turb. Gen. Start. Motor					C	160
23QT-1	Lube Oil tank heater					C	7,50
23QT-2	Lube Oil tank heater					C	7,50
23FG-1	Fuel Gas Electric Heater						60
88BA-1	Turbine enclosure Ventilation duty fan					C	13,50
88BA-2	Turbine enclosure Ventilation duty fan						13,50
88FC-1	Oil Cooler Fan Motor						3,10
88FC-2	Oil Cooler Fan Motor						3,10
88FC-3	Oil Cooler Fan Motor						3,10
23WK-1	Heater OFF-LINE washing skid						4,00
88TW-1	Water wash pump motor OFF-LINE skid						2,20
88QA	Aux. Lube Oil Pump Motor					D	10,00
88QV	Lube Oil vapour separator motor					C	1,50
DCP-A	Direct current supply panel side A					C	25,00
DCP-B	Direct current supply panel side B					C	6,50

Electrical Load List

Consumers are classified according their frequency of operation, as continuous, intermittent or spare.

E - "Continuous"; loads of machines or consumers which operate continuously when the plant is in operation, except for breakdowns.

F - "Intermittent "; machines or consumers with a start-stop cycle: pumping, storage, loading…

G - "Spare"; machines or consumers which act as a spare for other machines and which do not therefore normally operate when the plant is in operation.

Each type is assigned a coincidence factor, which is applied to its absorbed load to work out the total power requirement.

Intermittent consumers, such as offloading pumps working under start/stop cycle for instance, are counted 60%.

Electrical

Spare consumers, such as pump B that operates only in case pump A does not, are counted 10% only, etc.

The factored loads are summed up in the Electrical Load Summary, which gives the total plant power demand and also the load on each electrical equipment (switchboard, transformer) allowing its sizing.

Item	LV MCC MCC-002 (V=400/230V)		CONSUMED LOAD		
	Equipment No.	Description	Continuous (E)	Intermittent (F)	Spare (G)
			kW	kW	kW
1	LP003-1	Fire Fighting pump Bldg Light&Small Pwr	10,0		
2	LP003-2				
3	HSV-0011	Valve for gas metering station		1,3	
4	HSV-0012	Valve for gas metering station		1,3	
5	PM-032A	Fire Fighting Jockey Pump	5,2		
6	PM-032B	Fire Fighting Jockey Pump			5,2
Maximum of normal running plant load : kW = 16,7 (Est. 1·E + 0.6·F)			15,2	2,6	5,2
Peak Load kW = 17,2 (Est. 1·E + 0.6·F + 0.1·G)			Electrical Load Summary		

The most demanding operating modes, such as start-up of large motors, are considered to define the maximum load condition. This will size the power generation.

Maximum and minimum power requirements, and required availability, allow to define the number (redundancy) and capacity of power generators. A typical arrangement would include 4 generators, each having a capacity of 50% of the plant total power requirement. 3 generators will be running at 2/3 of their capacity while the 4th one could be under maintenance in normal circumstances. Should one generator trip, the remaining 2 will ramp up to full capacity, allowing no disruption in power supply, until the 3rd generator comes back on line.

Power supply to some consumers cannot be interrupted without impact on the production of the plant. Additionally, some consumers shall remain powered at all times to ensure equipment or plant safety: rotating machinery lube oil pumps, fire fighting water pumps, etc. These consumers are classified as "**essential consumers**", for which redundant power supply is required, on top of the power supply from the main power generators.

Back-up power supply is provided by diesel generators. Unlike the main power generators, which run on fuel (gas) fed from the Process, diesel generators have their own stand alone (diesel) fuel supply. In such a way, fuel supply is not dependent on plant operation. Sizing of the diesel generators takes into account the power requirement to re-start the main power generators, e.g., starters of gas turbines, etc.

The requisition for the main power generators and the diesel generators is prepared by the Mechanical Engineer. It includes the data sheet for the electrical part (alternator) prepared by the Electrical Engineer besides the data sheet for the driver.

All plant systems, i.e., Process Control System, Emergency Shutdown system, Electrical Control System, etc. shall remain operational in the event of loss of the power generation. Equipment of these systems are called "**vital**" and must remain powered at all times. An Un-interruptible Power System (UPS), with batteries, is provided for this purpose. Capacity of the UPS is the sum of the power consumption of the equipment of the above systems. This information is obtained from equipment vendor and can take time to finalize.

The architecture of the electrical distribution system is determined by a number of factors including:

- connection to external grid (On-Shore),
- voltage levels, which depends on consumers, e.g., large motor require MV instead of LV for ordinary motors, e.g., 11kV, 6.6kV, 400V, 230V, 110V DC, etc.,
- segregation between normal and essential consumers,
- number and location of transformers and Electrical sub-stations, which depend on the geographical distribution of consumers[1].

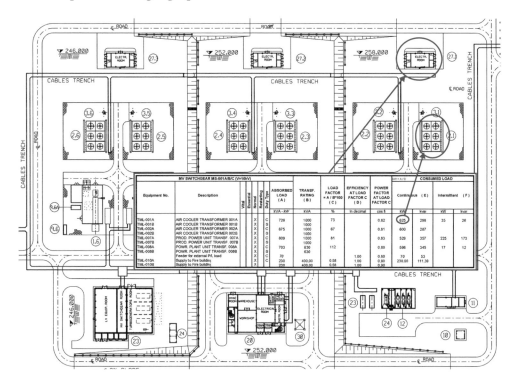

---

1. Sub-stations shall be as close as possible to main consumers to reduce cable length and section: on the plot plant shown here the power plant is item 23. Power supply to the gas-coolers (items 2.1-6), which are large low voltage consumers, is not done directly from the power plant but through sub-stations 27.1 to 3 equipped with high/low voltage power transformers. In such a way, high voltage cables are provided between the power plant and the sub-stations, which reduces the cable section, whereas low voltage cables, with large section, are required only on the short distance between the sub-stations and the consumers.

The overall power generation and distribution system is depicted on the **General One Line Diagram**, which shows generators, switchboards of various voltage levels, transformers and main consumers.

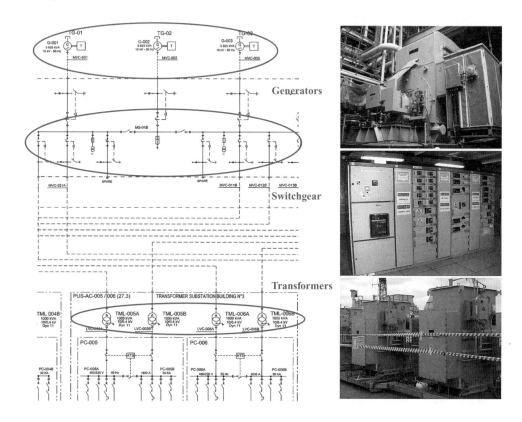

The Electrical Engineer specifies all equipment of the distribution system: switchboards, transformers, etc.

It produces a *data sheet* which, together with a *specification*, usually a general specification per type of equipment, will form the requisition for purchase.

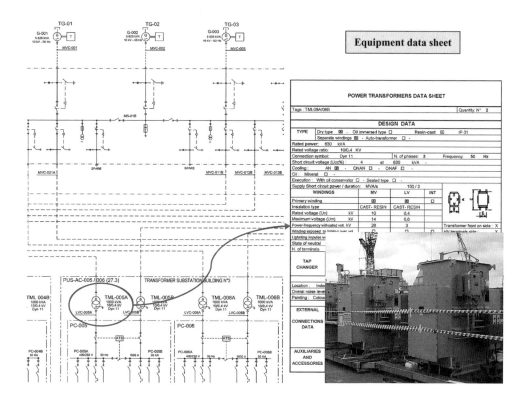

Electrical

Single Line Diagrams are produced for electrical switchboards, specifying to the vendor the content of the switchboard (incomers/outgoers), capacity, protections, control and monitoring devices.

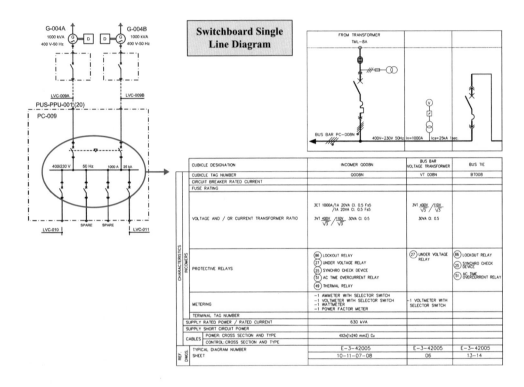

The power connection, the control, indication and remote monitoring features of switchgear cubicles are specified, for each type, e.g., motor outgoer, on the Switchgear Typical Diagrams.

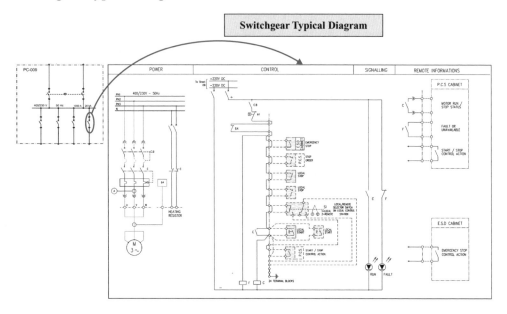

The electrical power distribution is monitored and controlled by an automated system: the Electrical Control System.

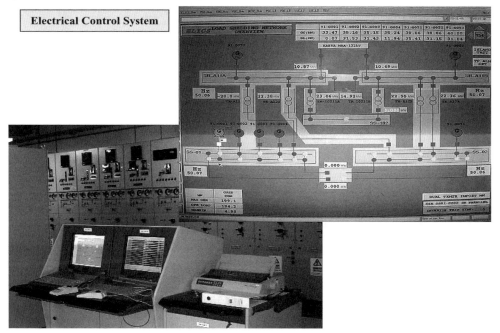

The Electrical Control System allows monitoring (status of protections, voltage/amperage/power values) at various points of the electrical system and control (start/stop of motor, etc.).

It also performs the key function of load shedding, interrupting power supply to non-essential consumers upon loss of power from the main generators, in order to reserve the limited power available, supplied by the emergency generators, to essential consumers. In the scheme shown here, for instance, the Automatic Transfer Switch will open the bus tie upon loss of normal power (from the main generators) in order to shed the non essential consumers, such as process pumps. The power supplied by the emergency generators is thus segregated and directed to essential consumers, connected to the right side of the bus bar, such as the fire water pumps.

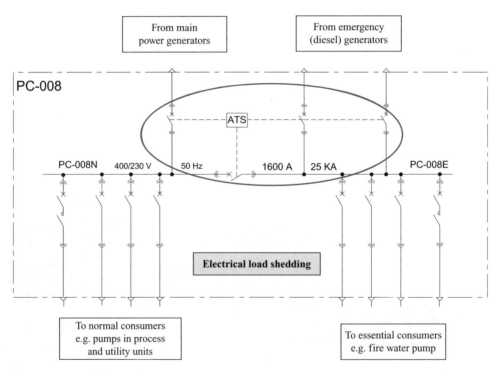

Electrical

The Electrical Control System is interfaced to the Process Control System, e.g., pump start/stop command is received from the PCS. It is also interfaced with the vendor supplied control system of the power generators.

A specification is produced to define the functionalities and capacity of the Electrical Control system: architecture and geographical distribution of equipment (allowing the vendor to identify the number and location of equipment its system will connect to, such as electrical switchboards, generator control equipment, etc.).

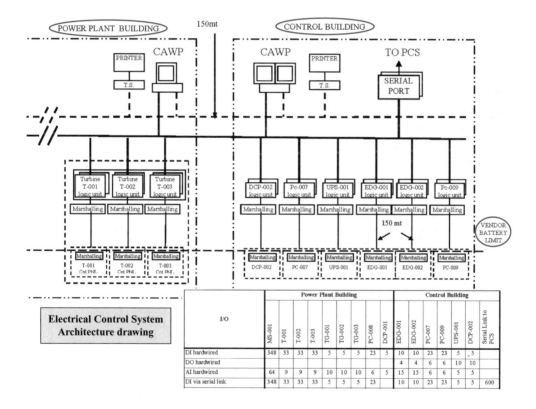

Electrical Control System Architecture drawing

I/O	Power Plant Building								Control Building						Serial Link to PCS
	MS-001	T-001	T-002	T-003	TG-001	TG-002	TG-003	PC-008	DCP-001	EDG-001	EDG-002	PC-007	PC-009	UPS-001	DCP-002	
DI hardwired	348	33	33	33	5	5	5	23	5	10	10	23	23	5	5	
DO hardwired										4	4	6	6	10	10	
AI hardwired	64	9	9	9	10	10	10	6	5	15	15	6	6	5	5	
DI via serial link	348	33	33	33	5	5	5	23		10	10	23	23	5	5	600

As discussed above, data from equipment vendors is not available initially, including power consumption which is estimated at first. When actual power consumption is known from vendor, the capacity of all electrical generation and distribution equipment is checked (main power generators, emergency generator, switchboards, transformers, cables, etc.).

Once the electrical equipment (switchboards, etc.) is purchased, its actual size will be known. This will allow the electrical engineer to define the equipment arrangement inside sub-stations, including provision for spare, which is shown on the Electrical equipment layout drawing.

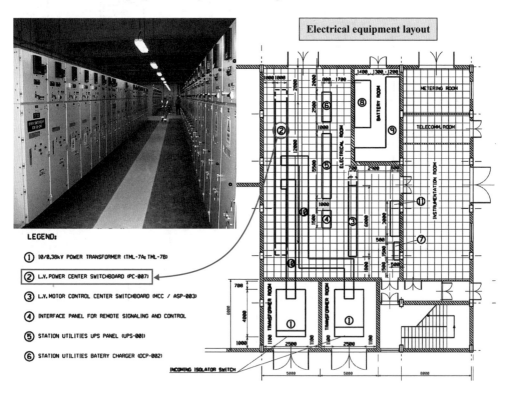

Electrical discipline also contributes to the specification of the mechanical equipment by preparing the electrical data sheets for the motors that are the drivers of these equipment, e.g., pumps, gas-coolers, etc. Such data sheet specifies in particular the type of explosion protection required for the equipment.

Indeed, an electrical field equipment located in an area where an explosive atmosphere can form shall have a special design so that it cannot be a source of ignition.

Such special design, called Hazardous area (Ex) classification of the equipment, is specified by the Safety engineer, according to the type of explosive atmosphere, its probability, ignition energy and temperature, etc. Refer to the Safety section for more details.

The Electrical engineer implements the Ex requirement for the various types of electrical equipment (electric motor, electrical socket, local control stations, etc.).

Electric cables are sized in order not to exceed a certain temperature under normal duty and also to sustain the short circuit current (the cable shall be able to handle short circuit of field equipment until the circuit breaker located at switchboard feeder opens). The Cable specification is governed by service, e.g., fire resistance for cables supplying critical equipment, armour for outdoor service, etc.

The specification of each cable is shown in the **Cable Schedule**.

Besides the electrical power distribution network, Electrical discipline also designs:

- the lighting system (as per illumination level requirements in each area),

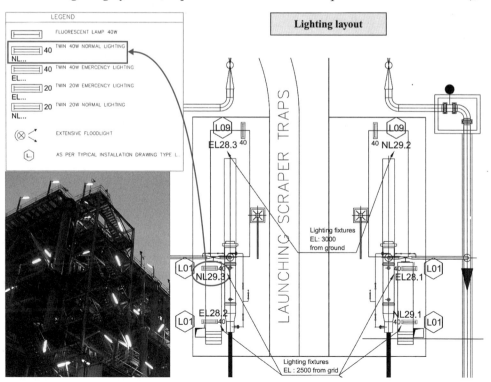

- the earthing system,
- the lightning protection system,
- the underground piping cathodic protection (see Material & Corrosion section),
- the heat tracing system of some process lines (to avoid freezing).

Electrical installation studies result in the production of all drawings required to install and connect the electrical equipment at Site:

- Electrical **cable routing drawings**,

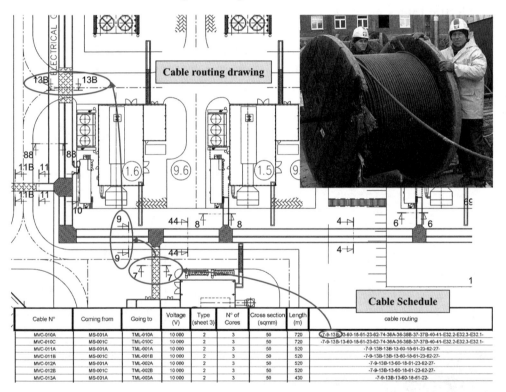

- Electrical **cable schedule** (showing the list of cables, the type, such as fire resistant, section size, number of cores, length of each cable),
- **Typical installation drawings**, for power, lighting, earthing, heat tracing, etc.,

Electrical

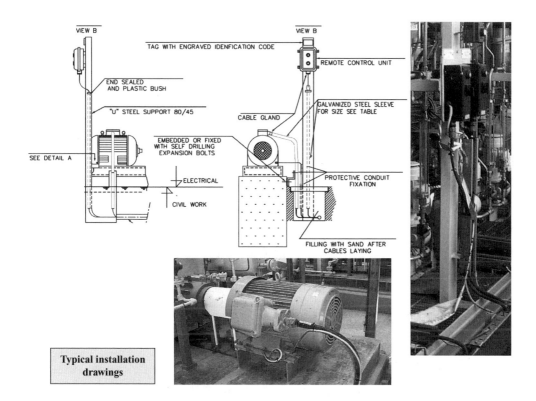

Typical installation drawings

- **Block diagrams** show typical (repetitive) connections,

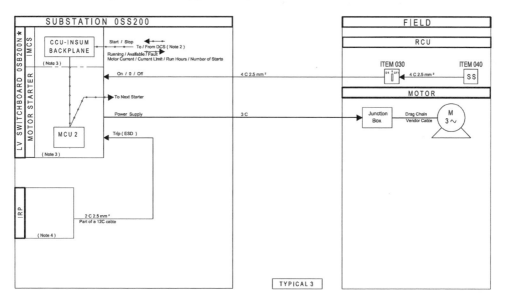

- **Electrical equipment location drawings**, showing location of all electrical consumers: motor local control stations, field sockets, lighting fixtures and junction boxes, etc.,

Alongside installation drawings the **Electrical bulk material take-off** is prepared in order to purchase cables, cable ladders, motor local control stations, junction boxes, cable glands and all other small installation materials.

ITEM	DESCRIPTION	QTY
1	local control station enclosure with: – 1 "START" push button with 1NO + 1 NC contact block – 1 "STOP" push button with 1NO + 1 NC contact block – 1 cable entry and metallic cable gland (non armoured cable 5 G1,5)	27
2	Welding socket 63 A – 400V – 3Ph + E – IP44 with: – connection to 35mm2 terminal – 1 cable gland for non armoured cable (4G35)	18

Lastly, Electrical discipline produces the **Trouble Shooting Diagrams**, which show the wiring of each consumer and will also be used for the Plant maintenance.

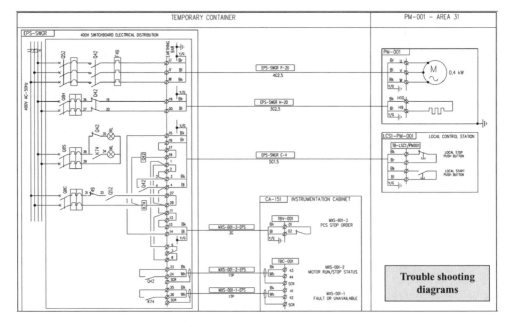

Trouble shooting diagrams

The electrical generation and distribution system is modelled using a computer software allowing to perform calculations and run simulations.

Simulations will include, for instance, the loss of one of the main power generators. The resulting transient conditions, before the stand-by generator has taken over, are checked to ensure that, for instance, process pumps will not have stopped.

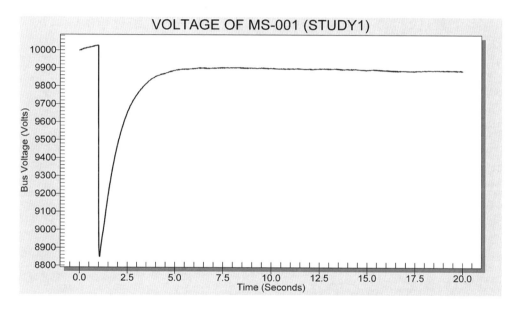

Final **Electrical calculations** are performed once all consumers and electrical equipment characteristics are known, all cables are sized, etc. The calculations will define the right setting of electrical protections. This right setting ensures selectivity. Selectivity means that, in case there is a short circuit on a motor, the protection of that motor only will open, no higher level protection will open, leaving the other consumers unaffected. The results are collected in the **Electrical Relay Schedule**, which is used at Site during commissioning to set the protections.

Chapter 13

Field Engineering

◆

The description above related to Engineering activities performed in the home office. When a Project goes in Construction phase, a small multi-disciplinary "Field Engineering" team made of engineers and draftsmen is seconded from the home office to the construction Site.

These Engineers and draftsmen are fully familiar with the engineering documents and drawings that have been produced.

They know on which document to find information.

Their first task is to familiarize the Construction contractor(s) working at site with the Engineering deliverables.

They are also there to solve issues discovered during construction, such as:

- engineering errors, such as interferences between a pipe and a steel structure,
- construction errors, e.g., a foundation has been cast slightly off its designed position and a design change is required to avoid re-cast,

- Site, equipment or material conditions differ from what was anticipated,
- overlooked engineering: the construction contractor needs some information that have not been prepared, e.g., cable routing was not defined in full, etc.,
- additions to the design. During the final inspection of the facility with the client before the hand-over a number of shortcomings are identified in the design, such as lack of access to valves as shown here…

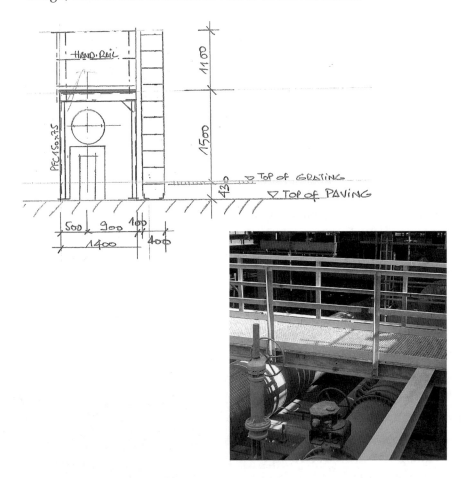

The Field Engineer performs the corresponding design. It would typically entail a survey of the location, dimensional measurements, sketching a solution on the spot, going back to the office to draft the drawings, issue the bill of material, etc.

Field Engineering

Changes to the design made at the Site must be approved by Engineering. To this end, the **Site Query** system is put in place:

Upon identification of a required change, the construction contractor issues a Site Query to the Engineer.

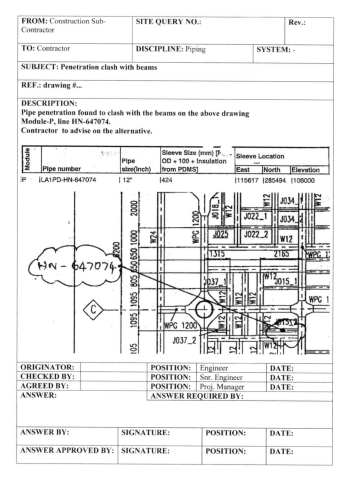

The Site query describes the issue encountered and, preferably, proposes a solution. The Engineer checks that the proposed change is acceptable or proposes an alternative.

In order to always work with up-to-date documents, Engineering updates a unique, called MASTER, set of engineering documents, with all changes. Changes are usually marked by hand and in red on the drawings, which are for this reason called "red-line mark-ups". The reference of the change is indicated next to the mark to trace it.

The Master set of **red-line mark-ups** is the reference on Site to which every party (Construction, Commissioning, etc.) refers.

							PESD 94-Y061A									
SITE MASTER			EFFECT				Item/tag	94-Y061A-P001A	94-Y061A-P001B	94-Y061A-P002A	94-Y061A-P002B	94-Y061A-P003A	94-Y061A-P003B	94-Y061A-P005A	94-Y061A-P005B	94-Y061A-X006	94-Y061A-P006A	94-Y061A-P006B	94-Y061A-X007
			1: immediate stop when alarm appears													
			1C: immediate close when alarm appears													
			1O: immediate open when alarm appears													
CAUSE			94UZ-6866A				DESCRIPTION	Seawater booster pump A motor	Seawater booster pump B motor	Distillate water pump A motor	Distillate water pump B motor	Brine pump A motor	Brine pump B motor	Antiscale pump A motor	Antiscale pump B motor	Antiscale mixer motor	Antifoam pump A motor	Antifoam pump B motor	Antifoam mixer motor
Rev			ALARM			Inhibition	Timer									
	TAG ALARM	SENSOR	Description	Signal												
4	94LAHH-6868A	94LIT-6868A	BR LVL CELL 5 VERY HIGH *HIGH*	Analog input	-		10 f s	1	1	1	1	1	1	1	1	1	1	1	1
4	94PAHH-6875A	94PIT-6875A	94Y061AX001 DSCH PRES O/S *HIGH HIGH*	Analog input				1	1	1	1	1	1	1	1	1	1	1	1
7	94FALL-6880A	94FIT-6880A	94Y061A SWTR CLG RJCT FL LOW *LOW*	Analog input			10 f s	1	1	1	1	1	1	1	1	1	1	1	1
7	94FALL-6895A	94FIT-6895A	94Y061A DW FL OUT B/L *LOW LOW*	Analog input			10 f s	1	1	1	1	1	1	1	1	1	1	1	1
7	94FALL-6896A	94FIT-6896A	94Y061A BR FL OUT B/L *LOW LOW*	Analog input			10 f s	1	1	1	1	1	1	1	1	1	1	1	1
4	94PALL-6884A	94PIT-6884A	94Y061A INSTRUMENT AIR VERY LOW LEVEL	Analog input	-		10 f s	1	1	1	1	1	1	1	1	1	1	1	1

At the end of the Project, red-line mark-ups allow to revise the engineering documents with all changes and issue a final "**As-Built**" revision. As-built's are part of the final documentation handed over to the client, and are used for the Plant Operation, maintenance, future expansion, etc.

Chapter 14

The challenges: matching the construction schedule

As explained above, Engineering delivery is twofold: the list and specifications of all equipment and materials, issued to Procurement for purchasing, and the construction drawings, issued to Site.

Ever decreasing EPC Project durations have stretched the engineering schedule: equipment must be purchased and construction drawings must be issued at a very early stage.

This has created a real challenge for Engineering.

Engineering deliverables (requisitions for ordering equipment and materials, drawings for construction) are indeed required to be issued in strict compliance with the Project schedule.

The logic of the Project schedule is the following: One starts by the end, i.e., required plant completion date, then works backwards, adding the duration of the various activities and their sequence, to work out the required start/completion date of each one.

In piping discipline, for instance, the Engineering, Procurement and Construction schedule will look as follows:

178   The challenges: matching the construction schedule

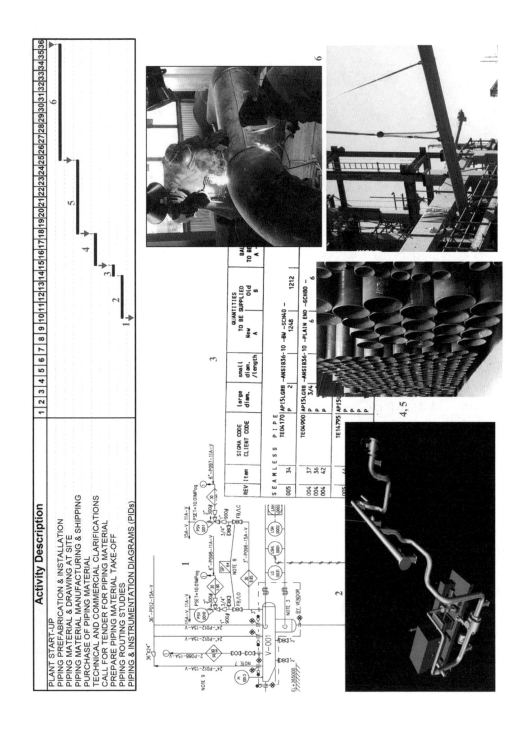

Which reads: Plant completion is due in month 36. If piping construction (pre-fabrication and erection) takes 10 months, then it should start at month 25. Piping material and drawings should be available at Site by this ROS (Required on Site) date. If piping takes 7 months to procure and ship, then it should be purchased by month 18 latest. Before Piping can be procured, inquiries must be issued to piping suppliers, then analyzed and clarified (technically, commercially). Allowing 3 months for the latter shows that inquiries shall be issued by month 15. Before inquiries can be issued, list of required material (Material Take Off) needs to be done, which takes 1 month: month 14. In order to do the piping material take-off, the piping routing studies must be completed, which takes 4 months, and must therefore be started on month 10. Piping routing studies are done on the basis of the P&IDs and Plot Plan, which are therefore required to be completed in month 10...

This retro-planning logic is how schedule requirements are defined for all activities of the project.

One sees that the schedule above, although dedicated to Piping, has set schedule requirements for P&IDs and Plot Plan, which are issued by other disciplines (Process resp. Plant Layout).

In fact, as depicted on the flowchart below, as piping studies go into higher levels of details (2D to 3D, etc.), they involve inputs from an increasing number of disciplines. The space occupied by the equipment/materials of all disciplines indeed needs to be considered to define the detailed pipe routes.

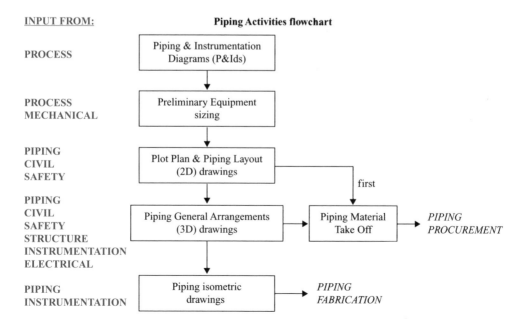

180   The challenges: matching the construction schedule

Engineering activities in the various disciplines are strongly dependent on one another, and on vendor information, as shown on the following Engineering activities and document flowchart.

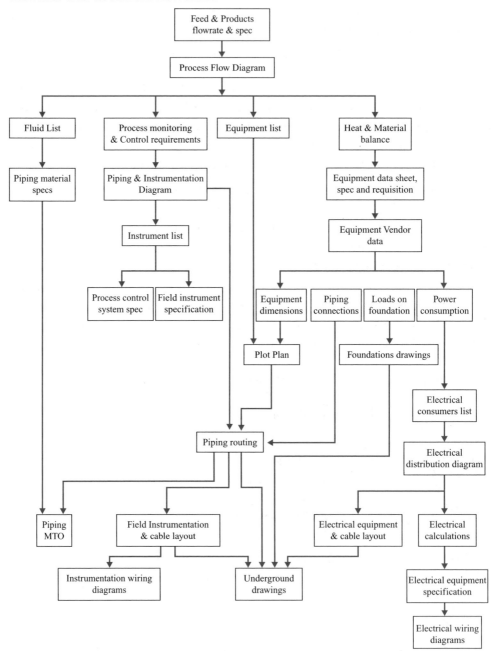

The challenges: matching the construction schedule

This translates into the main schedule inter-discipline relations shown here.

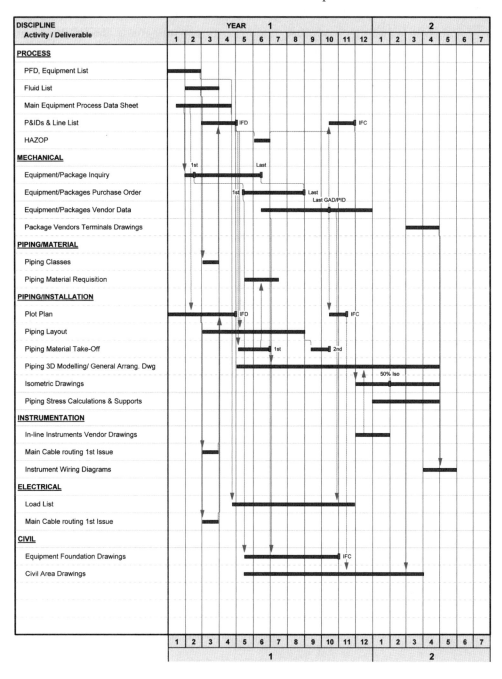

The challenges: matching the construction schedule

Integrating all schedule requirements and constraints results in an integrated schedule. The typical schedule of Engineering activities is shown in Appendix.

To meet the schedule, a number of short cuts, must be made. These short-cuts consist of making **estimates** before actual data is available. Such estimates must be as accurate as possible in order not to require rework later on. They must be conservative, i.e., contain some level of contingencies, but not too much in order to avoid a costly over-design.

In a lot of cases, engineering experience is required to anticipate a correct design before all information is available.

The design of a steel structure supporting pressure safety valves is one of them. The loads to be borne by the structure will depend on the position of these valves, as pressure safety valves will subject the structure to large reaction loads when operating. Such large horizontal loads at the top of the structure have a very significant impact on the required strength of the structure. The elevation of the safety valves must therefore be correctly defined as it is a key input to the design of the structure.

This elevation will depend on the routing of their inlet and outlet pipes. The routing must provide enough flexibility to allow for thermal expansion of these normally non flowing lines. This requires a routing with a number of direction changes, or even a purpose made expansion loop, which could determine a higher elevation for the valves.

Location and loads of pipe supports, which also serve as design input to the structure, must also be guessed. It is not be feasible to run calculations at this stage to confirm the envisaged pipe routing, as these calculations require too much details. Experience is required to define a proper routing, that will later be validated by calculations.

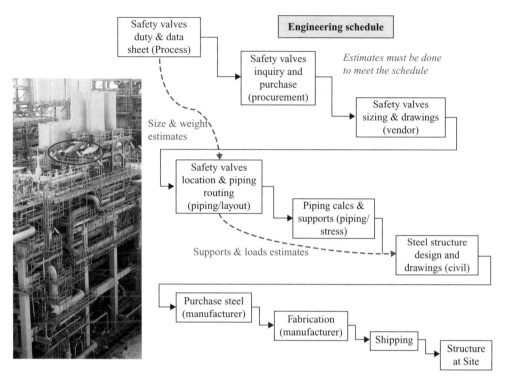

Engineering activities are highly dependent on availability of **vendor information**. Engineering indeed integrates individual vendor supplied equipment into an overall facility.

Layout activities for instance, depend on size information from equipment vendors. Such size will only be known once the vendor has completed its design. For complex packages, such as a turbo-compressor, the vendor will usually purchase sub units, such as the fuel gas unit, from sub-vendors.

This will further delay the availability of equipment dimensions and finalization of the layout and position of equipment.

This delay might put casting of the equipment foundation at Site on a very tight schedule.

Similarly, the electrical engineer might not be able to complete its design before the actual electrical consumption of the fuel gas heater is known. Timely availability of vendor information is critical not to impede the progress of the design. To this end, required submission dates of key vendor documents are specified in the requisition and penalties associated to delays in their submission.

The challenges: matching the construction schedule

Item	SUPPLIER'S DOCUMENTS – REQUIREMENT SCHEDULE	DOCUMENT STATUS	
	Designation	For approval	Approved final, comments incorporated
1	Dimensional outline drawings of turbocompressor set*	D + 45 days	D + 60 days
2	General arrangement DWG of turbocompressor building with inside and outside installations*	D + 45 days	D + 60 days
3	Air inlet and exhaust systems arrangement drawings*	D + 45 days	D + 60 days
4	Lube oil air cooler arrangement drawings*	D + 45 days	D + 60 days
5	Turbocompressor set foundation plan with static and dynamic loads*	D + 45 days	D + 60 days
6	Foundation plan with static and dynamic loads for turbocompressor building and other aux. equipment	D + 45 days	D + 60 days
7	Customer mechanical connections list and plan with max. allowable loads*	D + 45 days	D + 60 days

D : Effective date of purchase order

*Non-submittal of these documents within specified time delay will entail penalties

Ironically, the first drawings required at Site are produced last in the engineering work sequence. For example, foundations are one of the first activities done at Site, just after general earthworks. However, foundations data are the last data available for an equipment: first comes the equipment process duty, then its mechanical specification, then its design by the vendor, then only its size, loads, etc.

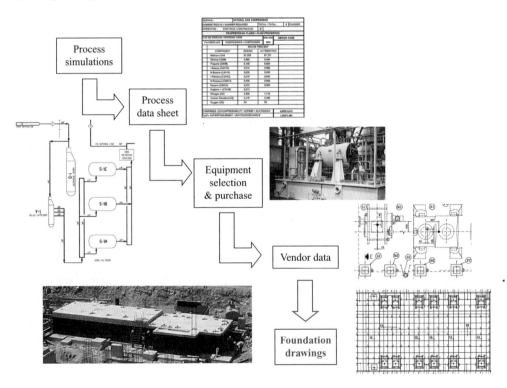

The challenges: matching the construction schedule

Schedule requirements for engineering deliveries related to specific equipment/networks will come from the **construction sequence**. Underground networks, for instance, must be installed very early. Indeed, their installation requires excavations which occupy a large footprint and prevent any overhead work (for safety reason). Hence, before above ground erection can proceed, undergrounds need to be installed and backfilled. This will require piping discipline to freeze the routing of underground piping and issue the material requisition early, and for electrical and instrumentation disciplines, whose cable networks are shallower hence installed later, to freeze the routing of cable trenches not so long afterwards, etc.

Another classical example is paving. Construction will aim to complete early the paving works in areas that will be crowded during erection activities, e.g., areas around equipment with a lot of connecting pipe work. It will indeed ensure a safer and more productive erection, avoiding underground/erection activities interferences. This will set an early schedule requirement to the Civil engineer to issue the paving drawings.

Another example is that of heavy equipment whose lifting requires close access of the crane to the equipment installation location. In such case, the crane may have to stand on an area to be built at a later stage during the installation of the equipment. Construction activities in such area cannot start before the equipment is installed. The installation of the equipment must therefore be done early, which will translate into a requirement for the civil engineer to issue the equipment foundations and supporting structures drawings at an early stage.

This is similar to what happens on an Off-Shore project, where the installation of large equipment requires clearance, which delays construction activities in the equipment installation way until after the equipment is installed.

Engineering deliveries to Construction do not only include the construction drawings. Ahead of production of such drawings, Engineering issues to Site the bill of quantities that allow the Construction contractor to plan its work. Combined with material delivery schedule, e.g., for pipes, it allows the construction contractor

The challenges: matching the construction schedule

to identify the required manpower and the appropriate time to mobilize it to avoid both idle time and shortage of manpower. This is particularly critical for an On-Shore project in a remote location where mobilization of resources takes time.

The challenges: controlling information

Engineering, unlike manufacturing, is not about processing matter but information, which is far more elusive. Managing information quality and flows is critical.

The nature of information varies, from technical requirements of the Client, design criteria, site data, output information from one discipline serving as input to another one, information from equipment suppliers, etc.

Parties exchanging information, either as a receiver or as a provider or both, are the Client, the various engineering disciplines, procurement, vendors, construction contractor(s), third parties, etc.

Procedures are produced as part of the **Engineering Quality Plan** and implemented to ensure quality of information and control over its exchange. These procedures describe in details how to implement the good practices, which this section will describe.

First of all, information is recorded and communicated in formal documents. Information is not communicated through email and other informal ways but through controlled documents, bearing a number, a revision, transmitted in a traced manner and listed in a register.

The first set of information is the one serving as input to the design: functional and performance requirements, design criteria, codes to be used, specifications to be followed, etc. It is defined and recorded in the **Design Basis** document.

The challenges: controlling information

Any information missing from the Client is requested in a recorded way through a **Query**. Any deviation from the contractual requirements will be submitted to the Client's approval in a formal and recorded manner by means of issuance of a **Deviation Request**. Technical information exchanged with third parties is recorded in **Interface Agreements**, which are described in more details at the end of this section.

Many engineering documents are not produced at once. Instead, they are developed progressively as the design progresses, with increasing levels of details, up-dated with information received from other disciplines, vendors, etc. The proper way to ensure that all modifications are properly collected in order to be incorporated in a document's next revision is to keep a document **Master** copy. All comments are collected on this one copy of the document that is clearly marked "MASTER".

Documents are issued in controlled manner by means of **revisions**. Document issues will be made for various purposes: internal review (a document from one discipline is distributed to the others for review), Client approval, issue for design (a document from one discipline is issued to serve as a basis for other disciplines), issue for construction, etc.

The purpose of the revision is usually indicated in the revision codification and label, e.g., IFA (Issue for Approval), IFC (Issue for Construction), etc.

A	23/11/2000	IFC- FIRST ISSUE
2	30/08/2000	UPDATED FURTHER NEW GENERAL ONE LINE DIAGRAM
1	16/06/00	UP DATED
0	29-06-99	ISSUE FOR CUSTOMER APPROVAL
Rev.	DATE	DESCRIPTION

Document revision is essential to the engineering process, where disciplines work from documents originating from other disciplines. When some information contained in a document needs to be communicated, issuing a revision of the document is the way to freeze the contained information in a certain stage. This freeze is essential to the receiving party, which cannot work with moving input data. Equally important is the definition, by the issuing party, of the validity of the data, and what its purpose is.

The challenges: controlling information

For instance, the cable list will be continuously up-dated throughout the design phase, with addition of cables, calculation of cable sections, definition of cable routings, cable lengths, etc.

The purchase of cables cannot, however, be delayed until the end of the design phase. The cable list will need to be issued to the purchasing department much earlier, consistently with the cable lead time (several months).

The revision of the cable list issued by Engineering to Procurement needs to clearly identify its purpose. The revision will, for instance, be labelled "issue for inquiry" or "issue for order 70%" (ordering 70% of quantities only will allow to prime the supply while avoiding surplus should cable length decrease as design progresses), etc.

Another example showing the importance of properly identifying the **document status** is the civil area, also called composite, drawings. These drawings show all underground constructions in a given area. Whereas main equipment foundations are defined early, other underground objects, such as cable trenches, pipe support foundations, etc. are defined much later. The area drawings being used to locate the main equipment foundation are needed early at Site. They must be issued at a time where pipe support have not been defined. They will therefore be issued initially with only part of the information – the main equipment foundation - valid. The revisions shall clearly specify the validity, e.g., rev 0 "for main equipment foundations", rev 1 "for foundations and underground piping", rev 2 "for foundations, underground piping, pipe support foundations", rev 3 "for foundations, underground piping, pipe support foundations, cables", etc.

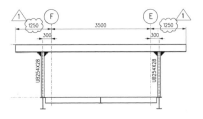

Revised parts of documents must be highlighted, which is usually done by means of "clouds". These revision marks allow the recipient of the revision of a document to visualize immediately what has changed compared to the previous revision, without having to read it all again.

It may happen that part of a drawing is not finalized at the time the drawing needs to be issued. In such a case, this part is highlighted (usually by means of an "inverted" cloud) and marked "HOLD".

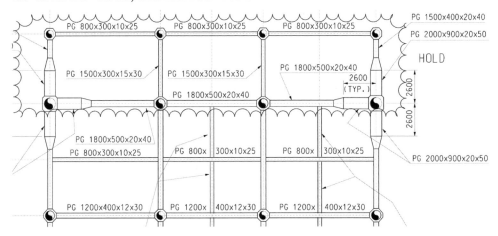

The design output must be checked before issue. Documents issued by a discipline are first checked within this discipline, by a person different from the one who prepared the document. The **verification** shall cover compliance to design basis, absence of errors, orders of magnitude checked, incorporation of latest changes, etc. Documents which concern several disciplines are submitted to an Inter-Discipline check. The discipline lead engineer determines which parties are to be involved in the review. The review can take the form of circulating the document to collect comments or by joint review during a meeting.

Each document is validated before it is issued. Such **validation** consist of checking that the document is fit for its purpose and complete, that the latest changes/instructions have been taken into account, etc.

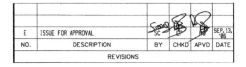

Evidence of the **verification** and **validation** (approval) is materialized by signatures on the document.

The challenges: controlling information

Information must be communicated to all concerned parties. A **Document Control Procedure** which includes a distribution matrix is prepared to ensure systematic distribution of each document according to type.

Issuing engineering discipline	Type of document	Document code	Safety and Environment	Process	Piping/layout	Piping/material	Piping/stress	Drafting office	Civil	HVAC	Structure	Electrical	Instrumentation & telecom	Vessels & Heat exchangers	Pipelines	Mechanical	Materials, coating, painting & welding	Cathodic protection
PIPING	LAYOUT DRAWING, PLOT PLAN	M1	x	x	x	x	x	x	x	x	x	x	x	x	x	x		x
	GENERAL ARRANGEMENTS DRAWING	M2	x	x	x		x	x	x		x	x	x	x	x	x		x
	LIST, MATERIAL TAKE-OFF	M4		x	x	x	x	x										
	ISOMETRIC DRAWING	M5			x			x										
	CALCULATION	M6			x		x	x										
	SPECIFICATION	M7			x	x	x	x					x				x	x
	DATA SHEET	M8			x	x	x	x					x				x	
	REQUISITION	M9			x	x		x										
INSTRUM	INSTALLATION DRAWINGS	A1	x		x				x	x	x		x	x				
	DETAILS DRAWINGS	A2			x				x		x		x					
	DIAGRAMS	A3			x				x				x					
	LIST, MATERIAL TAKE-OFF	A4	x	x	x				x				x	x				

This requirement translates in making sure that everybody uses the up-dated information hence has access to the latest versions of documents. The latest revision of documents is indicated in the **Engineering Document Register**.

Document number			Document title	Document revision
A	1	48104	Service building instrument. rooms cables routing	B
A	2	48102	Trouble shooting diagrams	D
A	3	48134	F&G system architecture drawing	E
A	4	50100	Instrument index	B
A	7	50003	Spec for instrument installation works and service	C
A	8	50960	Instrument Data sheets for temperature switches	B
A	9	50110	Requisition for pressure relief valves	B
M	1	62059	General plot plan	B
M	2	62020	Piping details standard	C
M	2	62070	Piping general arrangement Area 1	D
E	1	42020	Cable routing general layout	D

The challenges: controlling information

In the case where the document library is an on-line library rather than a paper one, this up-to-date requirement will be automatically met.

Quality of the design itself is challenged during **design reviews**. The design review is best done by people outside the project team who have "cold eyes" allowing them to stand back from the context and to question.

Consistency of the design at interfaces is checked during **Interface reviews**. Interface reviews are attended by the various engineering disciplines and focus on making sure that they all work on the same page.

In the longer run, experience gained on one job is fed back in the engineering company's method. Engineering guidance documents, templates and check list are produced/up-dated as a result of this **feed-back** process.

Mere copy/paste of documents from one project to the next puts at risk of over specifying by carrying over, without noticing, some constraints specific to one job to the other. Production of universal templates is preferable.

Information management is of paramount importance at Interfaces between parties.

Exchange of technical information takes place between numerous parties: between engineering disciplines, between engineering and vendors, between engineering and third party, e.g., contractors building other parts of the plant, etc.

The most numerous interfaces are **internal interfaces**: between engineering disciplines.

In many instances, the input information of one discipline is the output of another. Civil, for instance, designs structures supporting equipment and pipes. The civil design is directly dependent on the equipment and pipes to be supported: Piping defines to Civil the required geometry, and advises the location and values of the loads to be supported.

The challenges: controlling information

Such information must be precise and properly communicated. This avoids inefficient design such as the one shown here where the foundation of the pipe support is grossly oversized. This was due to the bad quality of the information provided by Piping to Civil.

The difficulty is that Piping will not have finalized its design when the information has, in order to comply with the schedule, to be given to Civil. Piping must therefore make some assumptions and include some allowances to avoid later changes.

Should a change occur, Piping should advise Civil, without delay in order to minimize the impact. In such a way Civil will be able to advise Site on time not to cast the foundation but to wait for a revised design.

Dedicated internal documents are issued from one discipline to the other for the exchange of information.

Timely exchange of these documents and precision of the information contained is key to ensure that the receiving disciplines works on the good information.

External interfaces include that with equipment vendors and third parties.

Proper interface with equipment vendors is key to the match, at Site, between equipment and their environment, e.g., piping connections, electrical and instrument connections, etc.

Review by Engineering of vendor documents and incorporation in the design is key to achieve this match.

Vendor documents are not finalized at once but will undergo revisions. Engineering drawings must be up-dated with revisions of vendor drawings.

On the picture shown here, the drawings of the thermo-well vendor has obviously not been reviewed by Piping which resulted in a mismatch at Site.

The flow of information between Engineering and vendors is not one way. The orientation of the nozzles of a pressure vessel, for instance, will be specified by Engineering, upon completion of piping studies, to the vendor. Gussets to be provided on the vessel for support of piping, platforms and ladders will also be defined by Engineering.

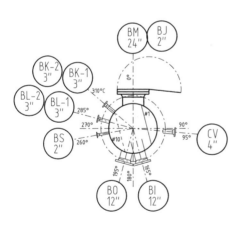

A number of interfaces with third parties are found on a Project, such as the ones found at the plant boundary: with the Contractor installing the inlet pipeline/outlet rundown lines, with Contractors in charge of other parts of the plant, such as the product tank farm, etc.

Information to be exchanged relates to the precise limit of supply of each party, and technical data at connecting point to ensure matching.

In the case of an interface on a pipeline, for instance, the technical data exchanged will not only include the coordinates of the connecting point, the type of connection (flanged/welded), but also more subtle data, such as the load (longitudinal force that could amount to several hundred kN) transferred from one side of the pipeline to the other.

The vehicle for the information exchange is the Interface Agreement, such as the one shown here.

		INTERFACE AGREEMENT		

Interface Agreement

Section/Step 1

Interface No:	Rev:	Page: of	Revision Date:
Title:			
Short Description:			Need Date:

Supplier:	Receiver:
Interface contact:	Interface contact:
Technical contact:	Technical contact:

Interface Details (Deliverable Description)

Discussion / Comments:

Reference documents and attachments:

Interface Agreement Approval

Supplier:	Date:	Receiver:	Date:
Printed Name:		Printed Name:	

Section/Step 2

Provided Interface Deliverable (description and document coding)

Interface Agreement Close-Out

Supplier:	Transmittal Date:	Receiver:	Closed-out Date:
Printed Name:		Printed Name:	

A proper Interface management must be put in place, such as one operating as follows:
- first, what interface information will be exchanged, who will be the giver and who will be the receiver and when the information will be provided is defined and formally agreed (step 1). Section 1 of the Interface Agreement is filled at this stage. This will allow the receiving party to plan its work. A proper definition and schedule of the information data is critical for the receiving party. Receipt of imprecise information or delay will indeed prevent the receiving party to proceed,

- second, the actual interface information, such as the one related to precise definition of battery limit shown here, is provided, usually by means of attaching an engineering drawing to the Interface Agreement. Section 2 of the Interface Agreement is filled at this stage.

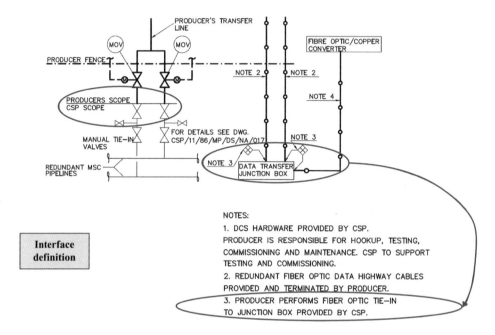

- the receiver finally confirms adequacy of the received data by the close out of the Interface Agreement.

Index: Common Engineering Documents

The list shown here is the standard list of engineering documents issued on a Project. An illustration of each one of them can be found at the indicated page.

General

    Engineering execution plan 6
    Engineering design basis 14
    Engineering quality plan 187
    Document control procedure 191
    Engineering document register 191

Process

    Process flow diagram 17
    Process equipment list 17
    Process fluids list 28
    Heat and material balance 18
    Process data sheet 19
    Process description and operating philosophy 20
    Piping and instrumentation diagram (P&ID) 21
    Process line list 29
    Process causes and effects diagram 23
    Emergency shutdown (ESD) cause and effect diagram 24
    Emergency shutdown (ESD) simplified diagram 25
    Emergency shutdown (ESD) logic diagram 25
    Emergency shutdown and depressurization philosophy 26
    Flare Report 26
    Calculation note 29
    Operating manual 30

Equipment/Mechanical

    Supply specification 36
    Mechanical data sheet 33
    Calculation sheet 32
    Vessel guide drawing 34

Index: Common Engineering Documents

Material requisition	36
Technical bid tabulation	37
Equipment summary	39

Plant Layout

General plot plan	43
Key plan	9
Unit plot plan	49

Health, Safety and Environment (HSE)

HAZOP report	53
Fire water demand calculation note	55
Fire water piping & instrument diagram	56
Deluge system arrangement drawing	57
Fire fighting equipment location drawing	57
Passive fire protection drawing	58
Fire and gas detection layout drawing	60
Fire and gas matrix	61
Hazardous area classification drawing	62
Quantitative risk analysis (QRA)	63
Health and environment requirements specification	68
Environmental impact assessment	69
Noise study	70

Civil (On-Shore)

Soil investigation specification	71
Grading plan	72
Foundation calculation note	76
Foundation drawing	77
Civil works specification	78
Civil standard drawing	78
Construction standard	79
Steel structure calculation note	84
Steel structure design drawing	84
Steel structure standard drawing	87
Civil works installation drawing	88

Building architectural drawing	90
Building detail drawing	91

Structure (Off-Shore)

Primary steel structure drawing	79
Structural steel specification	80
Steel structure material take-off	80
Secondary steel structure drawing	81
Weight report	40

Material & Corrosion

Material selection diagram	95
Painting specification	97
Insulation specification	98

Piping

Piping material classes specification	101
Line diagram	106
Piping layout drawing	110
Piping general arrangement drawing	112
Piping isometric drawing	113
Line list	116
Piping stress calculation note	118
Standard pipe support drawing	121
Special pipe support drawing	121
Piping material take-off	128

Handling

Handling equipment drawing	126

Instrumentation & Control

System architecture drawing	136
Functional diagram	134
Control narrative	135
Mimic display	137
Instrument index	140

Index: Common Engineering Documents

Instrument data sheet 141
Level sketch 142
Junction box location drawing... 144
Instrument location drawing 145
Cable routing drawing 146
Cable cross section drawing... 146
Cable schedule 146
Cable material take-off 147
Instrument hook-up drawing 147
Bulk material take-off 148
Standard installation drawing 149
Equipment layout drawing 149
Junction box wiring diagram 150
Loop diagram 151

Electrical

Electrical load list.. 156
Electrical load summary... 157
General one line diagram 160
Equipment data sheet... 161
Switchgear single line diagram 162
Typical diagram... 163
Control system architecture drawing..................................... 165
Equipment layout drawing 166
Lighting layout drawing 167
Cable routing drawing 168
Cable schedule 168
Typical installation drawing 169
Block diagram 169
Bulk material take-off 170
Trouble shooting diagram 170
Electrical calculation... 171

Appendix: Typical engineering schedule

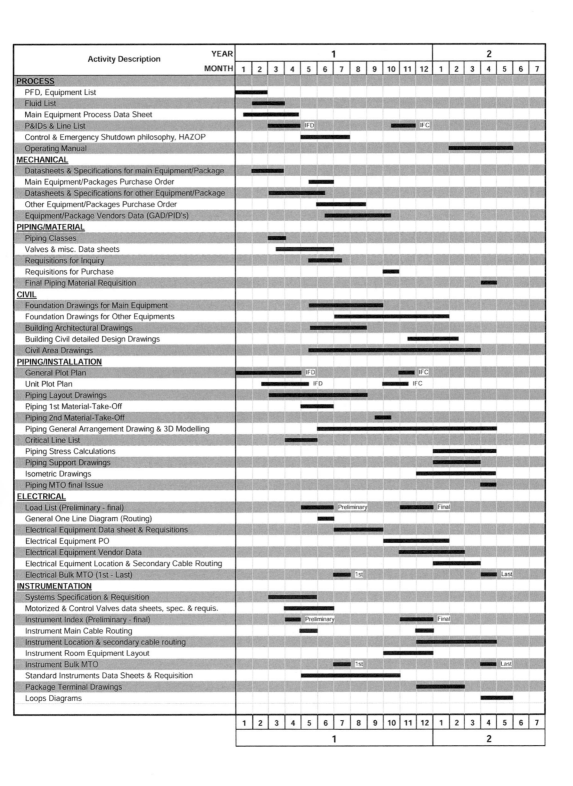

Achevé d'imprimer en septembre 2011 par EMD S.A.S. (France)
N° d'imprimeur : 25473 - Dépôt légal : avril 2010